Mass Spectrometry

Mass Spectrometry

James M. Thompson

Pan Stanford Publishing

Published by

Pan Stanford Publishing Pte. Ltd.
Penthouse Level, Suntec Tower 3
8 Temasek Boulevard
Singapore 038988

Email: editorial@panstanford.com
Web: www.panstanford.com

British Library Cataloguing-in-Publication Data
A catalogue record for this book is available from the British Library.

Mass Spectrometry

ISBN 978-981-4774-77-2 (Hardback)
ISBN 978-1-351-20715-7 (eBook)

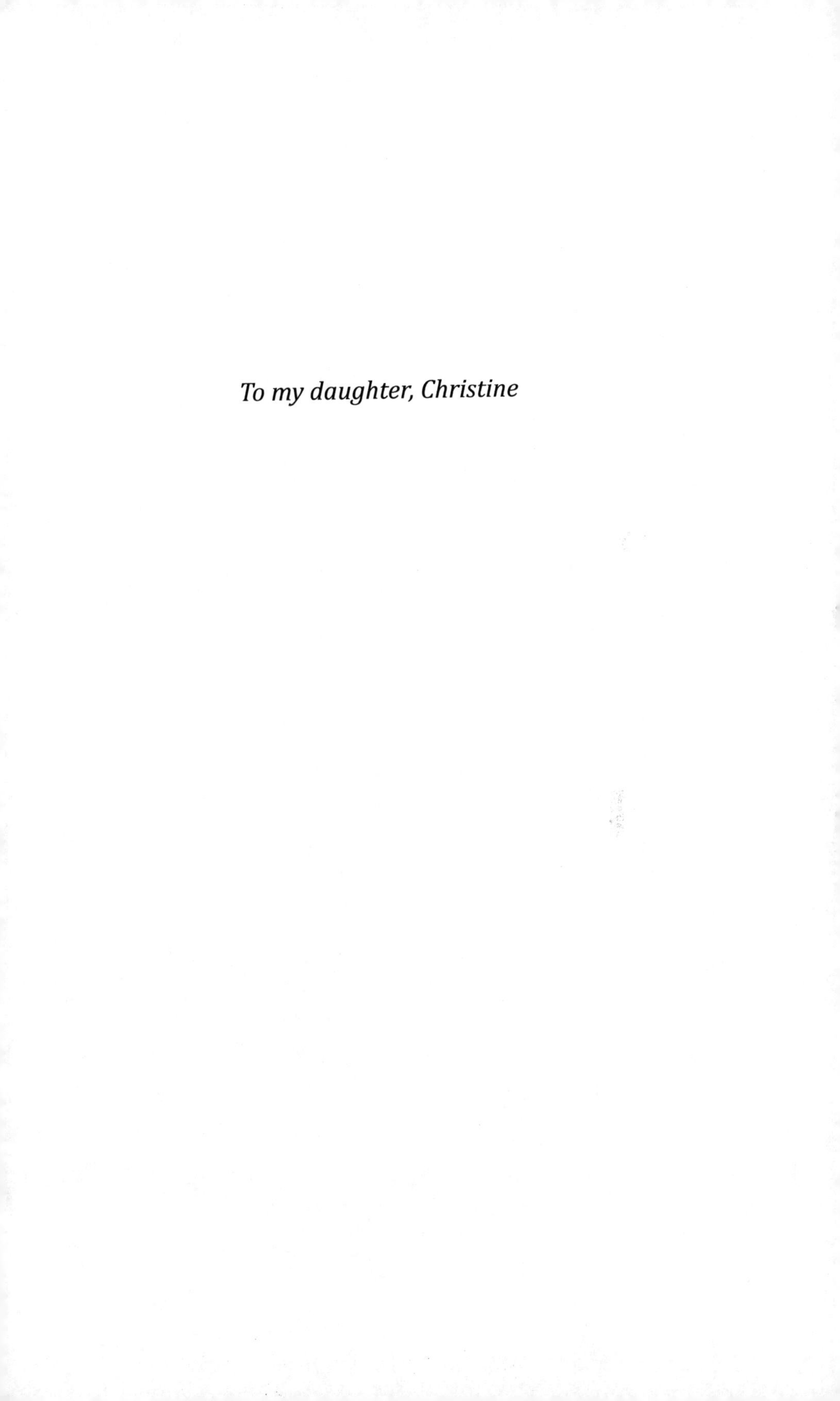

To my daughter, Christine

Contents

Preface

This is an introductory text, designed to acquaint undergraduate students with the basic theory and interpretative techniques of mass spectrometry. The author also believes that this material would be appropriate as an introductory text at the graduate level for those students lacking a background in the subject.

Much of the material in this text has been used over a period of several years in teaching portions of a one-semester course in materials characterization and chemical analysis. In addition to mass spectrometry, the course also includes a discussion of infrared spectroscopy and nuclear magnetic resonance spectroscopy. For the most past, the students in the course have had at least one year of organic chemistry. Thus, they have had at least a cursory exposure to the theory and interpretation of organic spectra.

In using this information in the course in materials characterization and chemical analysis, the author has devoted approximately 75 percent of the time to lecture discussions, with the remaining 25 percent devoted to the hands-on use of the various instruments at our disposal. However, depending upon the needs and assessment of the instructor, the text could serve solely as a lecture text for a one-semester course in mass spectrometry or it could be used to teach the mass spectra portion of a broader course in materials characterization and chemical analysis.

In this undertaking, the author has tried to put together a book that is readable at the undergraduate or beginning graduate level. To do this, the text has been interspersed with many illustrations, examples, problems, and an adequate bibliography. In developing an introductory text, out of necessity, many more advanced spectroscopic concepts have been excluded. Despite these exclusions, the text should serve to give the student a solid background in the area discussed. It is hoped that those using the text would be sufficiently inspired to continue with more advanced study in the area of mass spectrometry. Students learn best by examples. Accordingly, the book includes many examples of the spectra representing a broad range of organic compounds. Thus, the text has been made more readable.

No author is entirely satisfied with his or her final works. Even as the final draft of the text is reviewed, there is an inclination to go back and expand, rephrase, and modify certain sections. Since this effort could continue for some time, what has been done is reluctantly submitted with the recognition that there are many areas for improvement. In a desire to improve upon this material, reliance is placed upon you, the students and the professors. Your recommendations and notification of the errors of submission and omission will be appreciated.

James M. Thompson

Acknowledgments

The source of the mass spectra in this text has been the NIST/EPA/NIH Mass Spectral Database. The National Institute of Standards and Technology (NIST) has been very generous in allowing the use of their spectra. Also, the NIST has allowed us to reprint several tables that are listed in the appendixes. For this generosity, the author expresses his sincere appreciation.

Chapter 1

Some Fundamentals of Mass Spectrometry

1.1 Introduction

Mass spectrometry is an analytical-instrumental method, often used in association with other spectral data mainly for the purposes of determining the structures of organic compounds. The information obtained from the analyses of mass spectral data affords an excellent complement to IR, NMR, and UV spectra. The interpretation of the combined spectral data is often sufficient to establish the structure of the unknown compound. The mass spectrum of an unknown organic compound may also be computer searched against a large library of digitized mass spectra to establish structural identity. In this respect, the mass spectrum may be considered as the "fingerprint" of the molecule.

Unlike some other instrumental methods, organic samples subjected to mass spectra analysis suffer destructive fragmentation. However, a mass spectrum may be obtained with as little as a nanogram (10^{-9} gram) of sample material. Thus, the technique has an advantage over other instrumental methods in providing structure information on small and valuable samples. Mass spectrometry,

Mass Spectrometry
James M. Thompson

ISBN 978-981-4774-77-2 (Hardcover), 978-1-351-20715-7 (eBook)
www.panstanford.com

particularly when coupled with gas chromatography, is an excellent and highly popular technique for quantitative analysis and structural identification of sample mixtures. The availability of relatively inexpensive gas chromatographs/mass spectrometers has made it possible for small research laboratories and some undergraduate schools to acquire capability in this area. Nevertheless, high-resolution mass spectrometers, because of cost, are mainly limited to large universities and research institutions.

1.2 The Mass Spectrometer

In its most commonly practiced form, the technique of mass spectrometry involves bombarding vaporized organic molecules in a vacuum with high-energy electrons. This produces a variety of fragment ions of different masses which are separated and their masses and abundances measured. A simplified diagram of one type of mass spectrometer is shown in Fig. 1.1.

In obtaining a mass spectrum, the organic material is vaporized under a high vacuum in a heated chamber, after which it is leaked into an ionization chamber, where it is bombarded by a stream of high-energy electrons. The electrons are usually produced by either

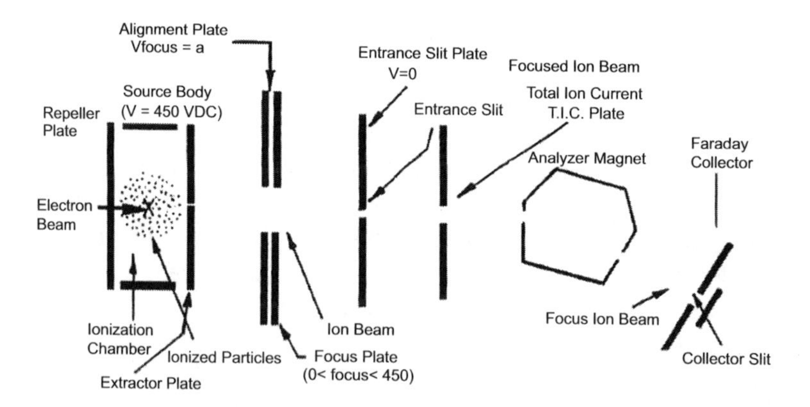

Figure 1.1 Schematic diagram of the major components of a magnetic sector mass spectrometer showing the ionization chamber, focusing system, and magnetic system. Courtesy of Varian Associates.

a heated rhenium or tungsten filament, and then energies may range between 10 and 100 electron volts (eV). Most mass spectrometers operate at 70 eV, since mass spectra are insensitive to small changes in electron energy in this range. The energy of the bombarding electrons (70 eV) is much greater than the ionization potential of organic molecules (which ranges between 8 and 15 eV). This relatively large energy difference is necessary if highly energetic ions are to be produced and spectra reproducibility is to be achieved.

Electron impact results in molecular destruction, producing a variety of fragment ions. Many of these fragments may be correlated with the structure of the original (parent) compound and their formation accounted for by known fragmentation patterns. Neutral, negatively, and positively charged fragments are produced, but the positive fragments are usually of major concern. Studies have shown that the fragmentation patterns of negative ions are similar to those of positive ions. The vast majority of the vaporized molecules escape fragmentation and are withdrawn from the chamber by a high vacuum.

1.2.1 The Magnetic Sector Mass Spectrometer

Regarding the magnetic sector mass spectrometers (Fig. 1.1), positively charged fragments are repelled out of the ionization chamber by a positively charged repeller plate (which has a potential difference between 800 and 8000 volts) into an acceleration plate. The large difference in potential across the accelerator plate increases the velocity of the positively charged ions, effectively converting their potential energy into kinetic energy, according to the following relationship:

$$\mathrm{eV} = \frac{1}{2}mv^2 \tag{1.1}$$

where Potential energy (eV) = Kinetic energy $\left(\frac{1}{2}mv^2\right)$

e = charge on fragment

V = acceleration voltage

eV = the potential energy of the ions in electron volts

v = the velocity of the ions

m = the mass of the ions

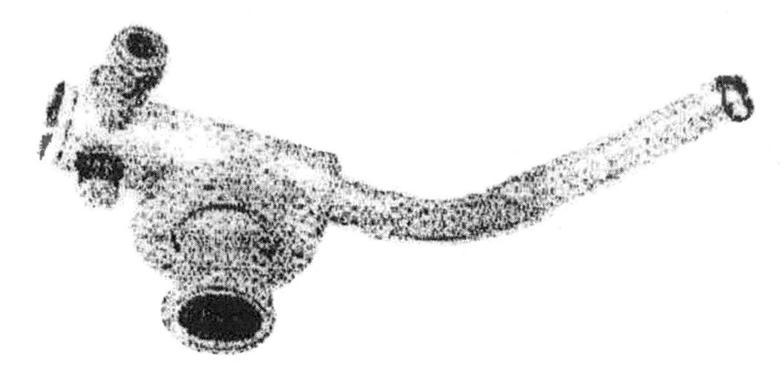

Figure 1.2 A photograph of an analyzer flight tube. Courtesy of Varian Associates.

After acceleration, the ions pass through a series of focusing slits, producing a uniform ion stream. They then enter the analyzer flight tube (Fig. 1.2) where they encounter a magnetic field. Most of the ions entering the magnetic field will have unit positive charge, while a few may be doubly charged (Section 1.9).

Once the positive fragment ions enter the magnetic field, they are deflected in a manner which depends upon their mass/charge (m/z) ratios. This deflection or curvature results in the separation of the ions into a series of "ion streams," with each individual stream consisting of numerous positive fragments of identical m/z ratios. The radius of curvature (r) of the ions is given by the following expression:

$$r = \frac{mv}{e\mathbf{B}} \tag{1.2}$$

where e is the charge on the ion, $\mathbf{B}$ the strength of the magnetic field in gauss, m mass of the ion, r the radius of curvature, and v the velocity of the ion.

By solving Eq. 1.2 for v (the velocity of the ion), and substituting into Eq. 1.1, the fundamental equation of mass spectrometry is obtained as shown in Eq. 1.3.

The fundamental equation of mass spectrometry is obtained as follows:

$$\frac{m}{e} = \frac{\mathbf{B}^2 r^2}{2V} = \frac{m}{z} \tag{1.3}$$

Considering the above equation, it is obvious that the radius of curvature (r) of ions in the magnetic field is directly proportional

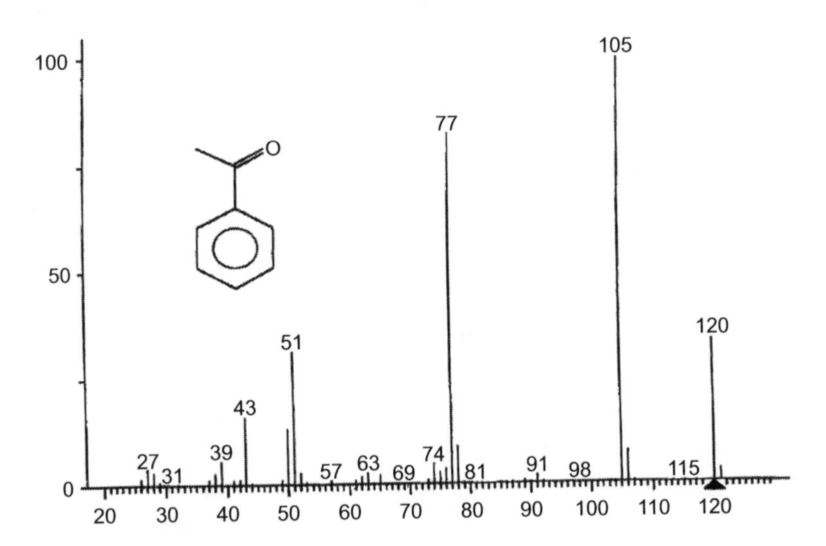

Figure 1.3 The mass spectrum of acetophenone.

to both the mass (m) of the fragment ion and its velocity (v) and inversely proportional to the strength of the magnetic field (**B**) and the charge on the ion (e). The radius of curvature of some of the ions are such that they will strike the analyzer tube. By changing the strength of the magnetic field and /or the potential across the accelerator plate, it is possible to direct each ion "stream" through a slit, located at the end of the analyzer tube. After entering the slit, the positive ions impinge upon a Faraday collector, electrons are removed, producing a current flow which is amplified and recorded as a line peak. The result is a spectrum consisting of an assortment of vertical lines (peaks or fragment ions) of different intensities, all appearing at their respective m/z values (Figs. 1.3 and 1.5).

In certain respects, the mass spectrum may be looked upon as the "fingerprint" of a molecule. In other words, if two mass spectra are identical and obtained under the same conditions using the same instrument, this may be taken as evidence of identical compounds or closely related isomers.

In a very simple analogy, if one breaks a glass object with a handful of rocks, numerous glass fragments will result. These fragments may be carefully arranged to recreate the original shape of the glass object. The object, before breaking, may be visualized as the neutral molecule, the rocks as the energetic electrons, and the broken pieces of glass as the fragments. In a rather general way, the concept of mass spectrometry is similar to the above analogy in that it involves the fragmentation of molecules and a recreation of the original structure from a study of the fragments produced. In other words, the fragments may be looked upon as pieces of a molecular jigsaw puzzle.

1.2.2 The Quadrupole Mass Spectrometer

The popular quadrupole mass spectrometer is characterized by a non-magnetic mass filter, which is composed of four solid rods, arranged symmetrically along the direction of ion flow (Fig. 1.4). To the quadrupole rods are applied both a radio frequency (RF) and a direct current voltage (DC). Once the positive ions are accelerated into the quadrupole field, they experience the influence of the applied radio frequency and direct current. Depending upon the RF/DC ratio and the m/z values, the ions will acquire either a stable or an unstable oscillation as they attempt to transverse the quadrupole field. Ions having the proper m/z values will have a stable oscillation and will successfully transverse the field and strike the detector. Ions with unstable oscillations will collide with the quadrupole rods and go undetected. It is possible to rapidly change the RF/DC ratio to scan a mass spectrum or to select specific RF/DC ratios to repeatedly measure a few specific ions (selected ion monitoring). Quadrupole mass spectral data are produced on a linear scale with equal spacing between ion masses. This feature facilitates recognition of the mass scale, even without a mass marker. The mass range achieved with the quadrupole mass spectrometer depends upon the energy of the radio frequency applied to the quadrupole rods. In some instruments, the output of the RF-generator is sufficient to allow scans up to 10,000 atomic mass units.

The DC voltage and radio frequency of quadrupole mass spectrometers can be rapidly changed while maintaining stable conditions throughout the mass region. This feature makes the quadrupole mass spectrometers ideally suited for computer interfacing.

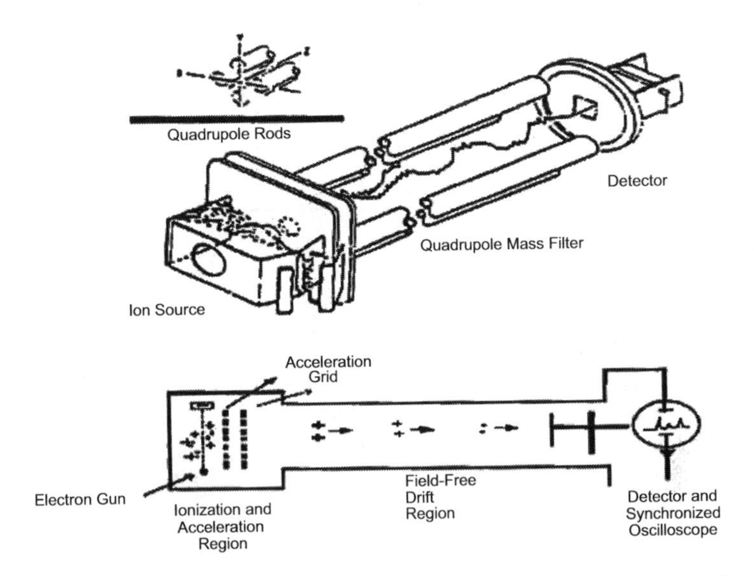

Figure 1.4 Schematic of the quadrupole mass filter, showing the oscillations of the positive ions (top) and the time-of-flight mass spectrometer (bottom). From Waller, G. R., Editor *Biochemical Application of Mass Spectrometry*, Wiley Interscience, New York, 1972.

1.2.3 The Time-of-Flight Mass Spectrometer

With the time-of-flight mass spectrometer (Fig. 1.4), the masses of the ions are determined by how long it takes them to reach the collectors. All fragment ions receive the same kinetic energy and are accelerated by a high-voltage accelerator plate. Although the ions will arrive at the collector in nanoseconds, their velocities will be different since they will have different masses. Therefore, they will arrive at the collector at different times. This results in their separation into groups according to their masses. The ions must be generated by a single pulse to start the ions simultaneously on their way. After the ions are recorded, the pulse can be repeated.

1.3 The Mass Spectrum

Each peak or fragment ion in a mass spectrum may be looked upon as a "piece" of the parent molecule, and the height of the

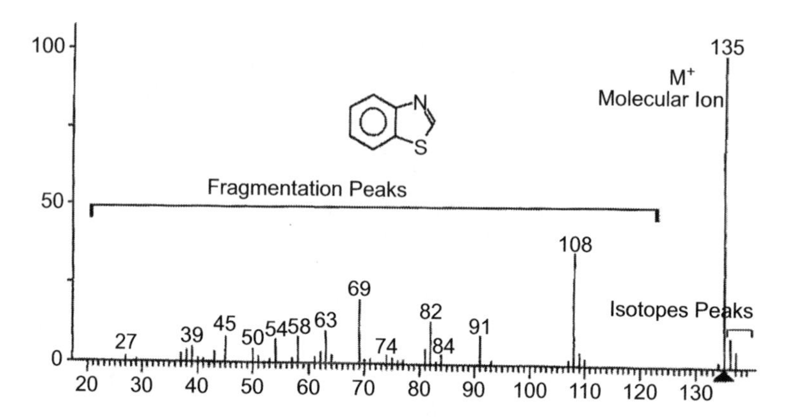

Figure 1.5 The mass spectrum of benzothiazole, showing some major features, including the isotopic peaks of carbon and sulfur at $(M + 1)^+$ and $(M + 2)^+$, respectively.

peak is directly proportional to the abundance of the fragment ion. Naturally, the more abundant fragments result from preferred modes of fragmentations. The major features of a mass spectrum are shown in Fig. 1.5.

1.3.1 Ion Fragments

Ion fragments may be produced by (a) simple fission, (b) rearrangement, and (c) elimination mechanisms. The correlation of fragment ions with structural features of the parent molecule is the basis of mass spectral analysis; however, some fragments cannot be correlated with the parent structure, since they may be produced by obscure rearrangement mechanisms.

1.3.2 The Base Peak

The base peak is always produced by the most preferred mode of fragmentation. It represents the most abundant fragment ion and is, therefore, the most intense peak in the spectrum. Usually a value of 100 units is assigned to the intensity of the base peak, and all other fragments are reported as a percentage of its intensity. In Figs. 1.3 and 1.5, the base peak is located at m/z 105 and 135, respectively.

1.3.3 The Molecular or Parent Ion (M^+)

The molecular or parent ion, M^+, is produced by the loss of one electron from the parent molecule and is therefore a radical ion. If present in the spectrum, the molecular ion is of major importance, since it represents the molecular mass of the compound. As shown in Fig. 1.5, it is not unusual for the base peak and the molecular ion peak to be one and the same. The M^+ ion appears in 75–80% of all mass spectra.

1.3.4 The $(M + 1)^+$ Ion

Another important mass spectral feature is the $(M + 1)^+$ ion. This ion usually represents the contribution of the ^{13}C isotope, which has a natural abundance of about 1.1%. This means that of all the carbons in the sample, 1.1% will represent the ^{13}C isotope. The intensity of $(M + 1)^+$ may be used, as we shall see, in approximating the number of carbons in the parent structure. In Fig. 1.5, the $(M + 1)^+$ peak is located at m/z 136.

1.3.5 Isotopic Fragments

As shown in Table 1.1, Cl and Br have two major isotopes which exist in fairly large natural abundance, with a mass difference of 2. As a result of the large natural abundance of these isotopes, compounds containing one or more Cl or Br atoms show pronounced $(M + 2)^+$ isotopic peaks. For instance, the spectrum, of methyl bromide (Fig. 1.15) shows two important fragments located at m/z 94 and 96, which represent M^+ and the $(M + 2)^+$ isotopic contributor. The two peaks have about equal intensities, reflecting the 50.7 and 49.3 natural abundance of the two bromine isotopes. For compounds containing one chlorine atom, the ratio of the intensities of M^+ and $(M + 2)^+$ should be approximately 3 to 1, again reflecting the 75.8 to 24.2 ratio of the natural abundances of the two major chlorine isotopes. Sulfur has four isotopes, but only two of the four have significant abundances. Because the ^{34}S isotope exists in about 4.2% natural abundance, compounds containing only one sulfur atom usually show a measurable $(M + 2)^+$ fragment whose intensity is approximately 4.2% of the intensity of the M^+ peak. In Fig. 1.5,

Table 1.1 Some isotopes and their natural abundances

Isotope	Natural abundance (%)
^{35}Cl	75.8
^{37}Cl	24.2
^{79}Br	50.7
^{81}Br	49.3
^{32}S	95.0
^{33}S	0.75
^{34}S	4.2
^{36}S	0.2

the $(M + 2)^+$ isotopic peak of sulfur is shown at m/z 137. As you may note from a careful measurement, the intensity of this peak is approximately 4% of the intensity of the molecular ion, located at m/z 135. For compounds containing two sulfur atoms, the expected intensity of $(M + 2)^+$ is approximately 8.4% of the intensity of the molecular ion peak.

Again, the ratio of the intensity of an isotopic peak to the intensity of its molecular ion, M^+, may be used to determine both the presence of, as well as the number of, certain atoms present in the parent molecule. (A detailed discussion of this matter is presented in later sections.)

1.3.6 High-Resolution Mass Spectrometry

Some of the more sophisticated mass spectrometers are capable of producing high-resolution mass spectra data which may be used to determine the elemental composition of fragment ions of similar masses. For instance, the ions shown below would give m/z values of 43 using a low-resolution instrument. However, with a high-resolution mass spectrometer, distinctly different masses would be obtained for each ion as shown.

$C_2H_3O^+$	43.0184
$C_3H_7^+$	43.0547
$C_2H_5N^+$	43.0421
$CH_3N_2^+$	43.0296
C_2F^+	42.9984

Table 1.2 The exact nuclidic masses of some common elements

Isotope	Atomic weight	Isotope	Atomic weight
^{1}H	1.00782522	^{19}F	18.9984046
^{2}H	2.01410222	^{28}Si	27.9769286
^{12}C	12.0000000	^{31}P	30.9737633
^{13}C	13.00335508	^{32}S	31.9720728
^{14}N	14.00307440	^{35}Cl	34.96885359
^{16}O	15.99491502	^{79}Br	78.9183326
^{18}O	17.99915996	^{127}I	126.9044755

Source: Wapstra A. H. and N. B. Gove. *J. Nuclear Data*, 9, 267 (1972).

High-resolution mass spectrometry is based on the knowledge that the atomic weights of the isotopes of elements are not whole numbers (Table 1.2). Each isotope is characterized by its unique "mass defect," which is related to the energy forces that bind neutrons and protons in the nucleus (or binding energy). For instance, the sum of the masses of protons and neutrons that makes up an atomic nucleus is more than the mass of the nucleus itself. This small mass difference is due to the binding energy of the atom and is related to Einstein's equation, $E = MC^2$.

As another example of the usefulness of high-resolution mass spectrometry, consider the fragment ions CO^+ and N_2^+. Using low-resolution mass spectrometry, both ions would give m/z values of 28; however, high-resolution mass spectrometry would be able to distinguish between the two. For instance, the exact mass of ^{16}O (based on the standard mass of 12.00000 for ^{12}C) is 15.99491, and for ^{14}N, it is 14.00307. Therefore, high-resolution mass spectrometry would result in an m/z value of 27.99491 for CO^+ and 28.00614 for N_2^+.

High-resolution mass spectrometers are capable of generating molecular masses to an accuracy of about 1 part in 100,000. Again, the advantage of this kind of accuracy is evident from a consideration of the two structures shown below for molecular mass 101.

$C_5H_9O_2^+$ m/z 101
$C_4H_5O_3^+$ m/z 101

Using low-resolution data, one could not distinguish between the two formulas; however, high-resolution mass spectrometer would be capable of making the distinction. For instance, based on Table 1.2, fragment ion $C_5H_9O_2$ would result in a molecular mass of 101.0687, while $C_4H_5O_3^+$ would give a value of 101.0238. Therefore, this deviation of pure isotopes from whole number masses, as we have seen, may be used to make distinctions among compounds of similar molecular masses.

1.4 Depicting Mass Spectral Data

As we have seen, mass spectra data may be depicted by a bar graph with the abscissa in m/z units and the ordinate in percent relative abundance. Another way of presenting this data is to tabulate the m/z values and their corresponding relative abundances. Table 1.3 and Fig. 1.6 show the tabulated data and bar graph for nitromethane. The tabulated data has the advantage of listing fragments whose relative abundances are too small to be observed on the bar graph. On the other hand, the bar graph gives a more visual picture of the fragmentation pattern, thus aiding in the interpretation.

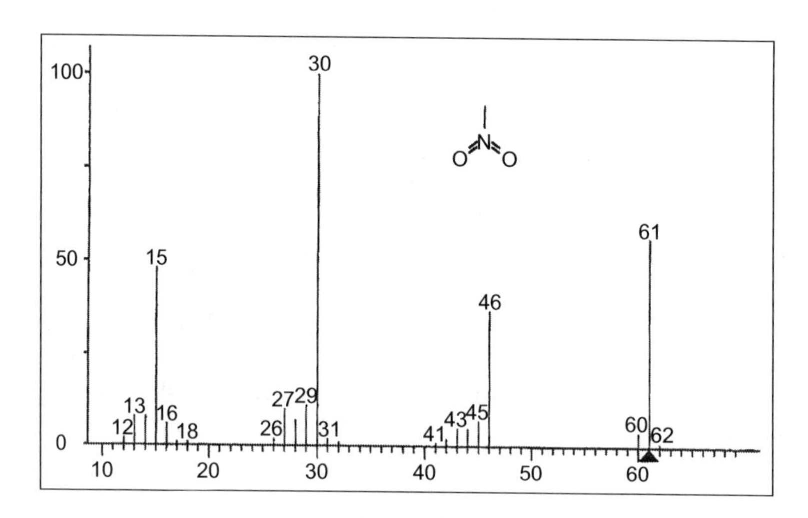

Figure 1.6 The mass spectrum of nitromethane.

Table 1.3 Tabulated data for nitromethane, showing the fragment masses and their relative abundances

Fragment	Abundance (%)	Fragment	Abundance (%)
12	1.9	42	1.5
13	3.5	43	3.1
14	8.3	44	4.7
15	51.0	45	6.2
16	5.6	46	35.0
26	1.6	47	0.17
27	6.9	48	0.15
28	6.3	59	0.06
29	8.0	60	4.0
30	100.0	61	53.0
31	1.4	62	0.92
32	0.53	63	0

1.5 The Molecular or Parent Ion

The molecular (or parent) ion, is usually present in the mass spectrum about 75–80% of the time. As previously mentioned, it is located in the high mass region and is produced by the loss of one electron from the parent molecule. Again, the importance of M^+ cannot be overemphasized since it represents the molecular mass of the compound.

The intensity of M^+ depends upon the structure of the parent compound. Sometimes it may constitute the base peak as shown in Fig. 1.5. More often than not, it is of low to medium intensity. In still other cases, such as for highly branched hydrocarbons, quaternary nitrogen compound and alcohols, the intensity of M^+ may be very small, or it may not be observed at all. Even if it is barely detectable, the identity of M^+ is just as valuable in determining the molecular mass of the sample compound.

1.6 Predicting the Formation of M^+

Under the conditions of electron impact, not all molecules will form molecular ions stable enough to be observed. Based upon a careful study of a large volume of mass spectral data and from a consideration of molecular factors such as bond strengths, ionization potentials, resonance stabilization, and hyperconjugation, it is often possible to predict the disposition of organic molecules toward molecular ion formation. Listed below are several statements that will aid in making these determinations:

(1) The molecular ion intensities increase in the following order:

Increasing M^+ Intensity ↓

- Alcohols
- Branched Hydrocarbons
- Carboxylic Acids
- Ethers
- Amines
- Ketones
- Mercaptans
- Unbranched Hydrocarbons
- Sulfides
- Alicyclic Compounds
- Conjugated Olefins
- Aromatic Compounds

(2) Non-branched, low molecular mass alkanes represent small targets and are likely to produce more intense molecular ions than their higher molecular mass counterparts.

(3) The molecular ion is more intense for unbranched cycloalkanes than it is for straight-chain alkanes. This results because cleavage of one bond in cycloalkanes keeps the mass of the molecule intact, while cleavage of one bond in straight-chain alkanes results in two lower-molecular-mass fragments.

(4) Branched alkanes are less likely to produce molecular ions than unbranched alkanes, since fragmentation at the point of branching produces relatively stable ions.

(5) Unsaturated hydrocarbons usually produce more intense molecular ions than saturated hydrocarbons because of the lower ionization potential of pi electrons compared to sigma electrons.

(6) Compounds containing allylic groups such as $RCH_2CH{=}CHR'$ will undergo fragmentation to produce allylic ions. These ions are stabilized by resonance, thereby reducing the intensity of the molecular ion. For instance,

$$R{-}CH_2{-}CH{=}CHR' \xrightarrow{-e^-} R{-}CH_2{-}CH\overset{\cdot +}{-}CHR' \longrightarrow \cdot R + CH_2{=}CH{-}\overset{+}{C}HR'$$

(7) Aromatic and heteroaromatic compounds (without alkyl side chains) usually produce observable molecular ions as a consequence of resonance stabilization.

(8) When heteroatoms such as N, S, and O are present in non-aromatic compounds, the intensity of the molecular ion is low or not observed at all.

(9) Compounds containing conjugated double bonds, because of resonance stabilization, often produce more intense molecular ions than those containing isolated double bonds.

(10) Mercaptans (RSH), as a general rule, show more intense molecular ions than alcohols or amines with similar structures.

(11) The probability of producing a molecular ion is significantly diminished if the molecule is capable of losing a neutral fragment through decomposition.

1.7 The Nitrogen Rule

Sample compounds having molecular ions with odd mass numbers must contain one nitrogen or an odd number of nitrogen atoms (Figs. 1.5 and 1.6). If the compound is known to contain nitrogen, then an even molecular mass ion means that the compound has an even number of nitrogen atoms. Exceptions to the nitrogen rule include NO and NO_2.

1.8 Metastable Ions

The vast majority of positive ions are accelerated through the magnetic field of the spectrometer and will be intact when they strike the collector plate. A few of the relatively unstable ions will be accelerated but will decompose within the analyzer tube, forming a smaller positive fragment (called a metastable ion) and a neutral fragment. The metastable ion will have a different radius of curvature and a smaller energy than the original fragment (since some of the energy is transferred to the neutral fragment). This secondary fragmentation process may be illustrated as follows:

$$\mathrm{m_1}^+ \longrightarrow \mathrm{m_2}^+ + (\mathrm{m_1} - \mathrm{m_2}).$$

In the example above $\mathrm{m_1}^+$ and $\mathrm{m_2}^+$ are the parent and daughter fragments, respectively. The observed mass m^* of the daughter ion is related to the true mass of the daughter ion, and the mass of the parent ion by the following expression:

$$m^* = \frac{m_2^2}{m_1}$$

Metastable ions are generally produced by fragments having lifetimes of less than 10^{-6} sec. They appear in the spectrum as broad, low-intensity peaks with non-integer masses, therefore, they are easy to recognize as shown by the example in Fig. 1.7. However, it should be noted that not all mass spectrometers are capable of detecting metastable ions.

Metastable ions can be important in confirming the existence of a parent-daughter ion relationship.

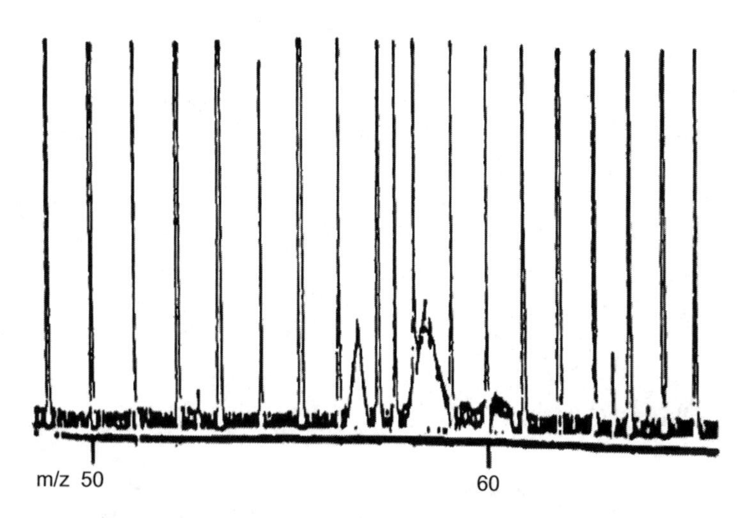

Figure 1.7 An enlarged portion of a mass spectrum, showing the broad, low-intensity metastable ions at m/z 56.5 and 58.5 (note the abundant doubly charged ion at m/z 57.5). *Source*: Johnstone R. A. W., *Mass Spectrometry for Organic Chemist*, Cambridge University Press, London, 1972.

1.9 Doubly Charged Ions

As previously mentioned, the majority of the ions striking the Faraday collector have unit positive charges and, therefore, their masses are identical to their m/z values. For some molecules, mainly those with available pi electrons such as olefins, aromatics and some heterocyclic compounds, there is a tendency to produce doubly charged ions.

Doubly charged ions appear in the spectrum at $m/2$. Hence, a doubly charged ion fragment appearing at m/z 80 actually has a mass of 160. Fragments having odd molecular masses will produce doubly charged ions with non-integer masses and are, therefore, easy to detect. As an example, the doubly charged ion located at m/z 57.5 in Fig. 1.7 stands out among the others. This ion actually represents a mass of 115. If a doubly charged ion is produced from an even mass ion, then its mass must also be even. In these cases, recognition of the ion as having a double charge is more difficult. However, doubly charged ions that are produced from even masses may show an $(M + 1)^+$ isotopic peak due to ^{13}C. If present, the $(M + 1)^+$ fragment should be conspicuous since it would be located 1/2 mass units beyond the related doubly charged ion, thus verifying its presence. It should be noted, however, that the intensities of doubly charged ions are usually relatively low when compared to singly charged ions.

1.10 The General Fragmentation Process

The fragmentation patterns of molecules under high-energy electron impact are not always completely understood and many of the fragments defy logical mechanistic explanations. This is especially true for those processes involving rearrangement mechanisms. Also, the fragmentation patterns of a variety of structural classes have not been studied sufficiently for characterization. Among these are many of the complex and high molecular mass compounds. There are, however, many well-established fragmentation patterns that, in some respects, resemble common organic processes such as pyrolysis, photolysis, and radiolysis. The fragmentation patterns of the common structural classes of organic compounds, i.e., ethers, alcohols, ketones amines, etc., have been thoroughly studied, and to a large extent, their fragmentation modes can be predicted with reasonable accuracy. Fragmentation of organic molecules under high-energy electron impact may be classified in a general fashion by one of three methods: homolytic, heterolytic, and hemiheterolytic fission.

Some general fragmentation modes

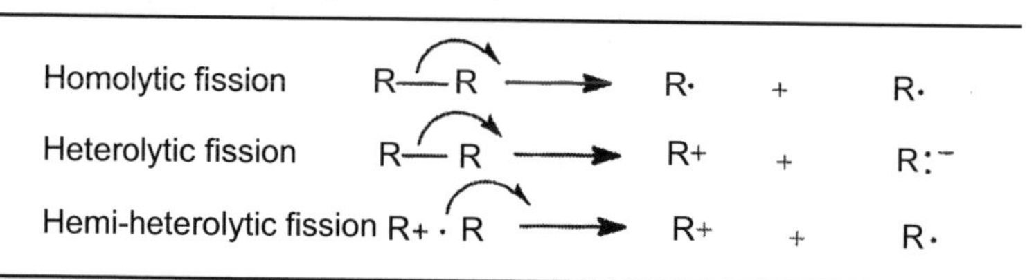

Note: A single electron movement is indicated by a "fishhook" arrow, while a normal arrow indicates the movement of two electrons.

1.11 The Fragmentation of a Hypothetical Molecule

The fragmentation patterns of molecules subjected to electron bombardment can be rather complex. The fragmentation process may include secondary fragmentation, ion rearrangement, and in some cases, ion-molecule interaction. The specific mode of fragmentation depends upon the structure of the sample molecules and involves, to varying degrees, molecular factors such as bond energies, ionization potentials, and molecular branching, as well as factors which stabilize the fragment ions, such as resonance and hyper-conjugation. A better understanding of the fragmentation process may be achieved by a study of the series of events illustrated below, using the hypothetical molecule (I). However, it should be clear that the events listed below do not fully convey the complexity of the fragmentation process.

In the examples below, fragmentation is usually initiated by the loss of an electron from a neutral molecule (reaction A) to produce the molecular ion, II, which is a radical ion. In cases where the molecular ion is relatively stable, its abundance will be relatively high. Depending upon the structure of the sample compound, the molecular ion may be relatively unstable, and under these conditions, it will undergo rapid decomposition and will either be unobserved or of low intensity.

The molecular ion, II, may undergo several different fragmentation processes to give positive ions and neutral radicals (reactions B, C, and D). As previously mentioned, usually the positive ions are expelled out of the chamber and detected by the spectrometer. Ion fragments consisting of two or more atoms may undergo additional fragmentation to produce still other ion fragments. Such a process is illustrated by the decomposition of ions IV and V into VII, VIII, IX, and X.

E-F-G-K (I) —A, $-e^-$→ $[\text{E–F–G–K}]^{+\bullet}$ (II)

B: $[\text{E–F–G–K}]^{+\bullet}$ (II) → E^+ (III, etc.) + F–G–K•

E (Rearrangement): $[\text{E–F–G–K}]^{+\bullet}$ (II) → F–G• (etc.) + E–K^+ (VI)

C: $[\text{E–F–G–K}]^{+\bullet}$ (II) → E–F^+ (IV) + G–K•

D: $[\text{E–F–G–K}]^{+\bullet}$ (II) → E–F• (etc.) + G–K^+ (V)

E–F^+ (IV) → E^+ (VII) + F

E–F^+ (IV) → E + F^+ (VIII)

G–K^+ (V) → G^+ (IX) + K

G–K^+ (V) → G + K^+ (X)

Also:

2 $[\text{E–F–G–K}]^{+\bullet}$ (II) + 2 E–F–G–K —Molecular-ion Combination (J)→ $[\text{E–F–G–K} \cdot \text{E–F–G–K}]^+$ (IX) + F–G–K• + $[\text{E–F–G–K–E}]^+$ (XII)

In reaction E, fragment VI is formed by a rearrangement mechanism. Fragments of this type often complicate the analysis by producing ions which appear to have no correlation with the parent molecule. For instance, the presence of m/z 29 ($C_2H_5^+$) in the spectrum of 2-methylpropane (Fig. 1.8) is an example of such a rearranged fragment.

Neutral molecules and fragment ions may collide during the fragmentation process. Although somewhat rare, the collision could

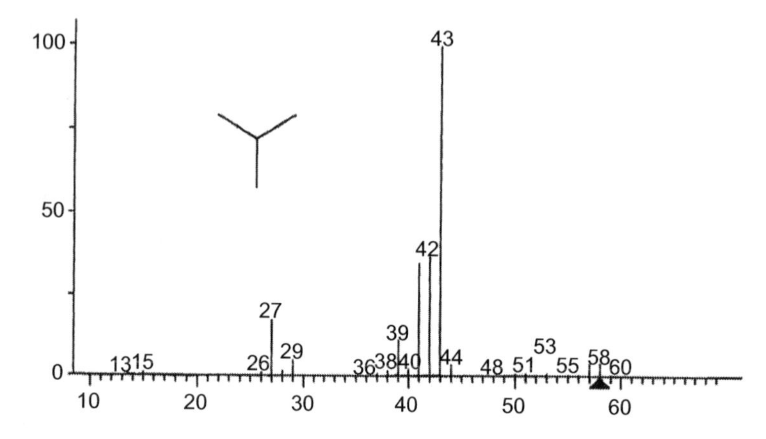

Figure 1.8 The spectrum of 2-methylpropane (isobutane) showing the rearranged $C_2H_5^+$ fragment at m/z 29, which is formed by an obscure rearrangement process.

produce ion-molecule fragments as shown by equation J. In this example, an ion fragment combines with a neutral molecule through a collision mechanism, producing an ion fragment whose molecular mass is higher than either of the colliding fragments as shown by fragment XI. Such occurrences are infrequent and when they do occur, ions of very low abundances are produced. Under certain conditions $(M + H)^+$ ions may be produced. This is discussed in Section 1.13.

1.12 Identifying the Molecular Ion

Of all the ion fragments, the molecular ion, M^+, requires the least energy for its formation. All other ions may be considered as originating from M^+ through some fragmentation process. The identification of M^+ is not always routine, for it does not always follow that the highest mass in the spectrum is the molecular ion. There may be several ions observed in the high mass region of the spectrum, and neither may correspond to M^+. In these cases, it is easy to mislabel the molecular ion or conclude that it is present when in fact it may not be. In cases where M^+ is barely detectable, its intensity may be enhanced either by increasing the sample size and/or by reducing the energy of the bombarding electrons. These procedures may also cause M^+ to appear in cases where it is initially absent.

If several fragments are grouped together in the high mass region of the spectrum, they may correspond to an isotopic cluster. These clusters are most obvious in the spectra of compounds containing bromine and/or chlorine and somewhat less obvious among compounds containing sulfur (see Figs. 1.5, 1.15 and 1.19). In still other cases, what may appear to be the molecular ion, may instead be an $(M + H)^+$ or $(M - H)^+$ ion, produced by a chemical ionization technique (Section 1.13).

1.13 Chemical Ionization

Chemical ionization (or CI) is based on the knowledge that under certain specific conditions some molecules may be selectively ionized to form $(M + H)^+$ and $(M - H)^+$ ions through ion-molecule collisions. The procedure is useful in determining the molecular weight (as well as other structural features) of unknown compounds that normally do not produce detectable molecular ions by electron ionization. To produce abundant $(M + H)^+$ or $(M - H)^+$ ions from sample molecules, we first take methane or some structurally related hydrocarbon gas and subject it to electron ionization in a special ion source at high pressures ($\sim$100 Pa) to produce the CH_5^+ "reagent" ions. When the unknown sample molecules are added to the environment of the CH_5^+ ions, collisions between "reagent" ions and sample molecules take place producing abundant $(M + H)^+$ or $(M - H)^+$ ions, depending upon the structural class of the unknown compound. Under the conditions required for chemical ionization, compounds containing a heteroatom such as ethers, esters, amines, aminoesters, nitriles, etc., will usually form abundant $(M + H)^+$ ions while saturated hydrocarbons will usually form abundant $(M - H)^+$ ions. Obviously, if these ions can be produced in abundance, the molecular weight of the unknown sample may be determined.

An example of the usefulness of the CI method is shown by comparing the electron ionization and chemical ionization spectra

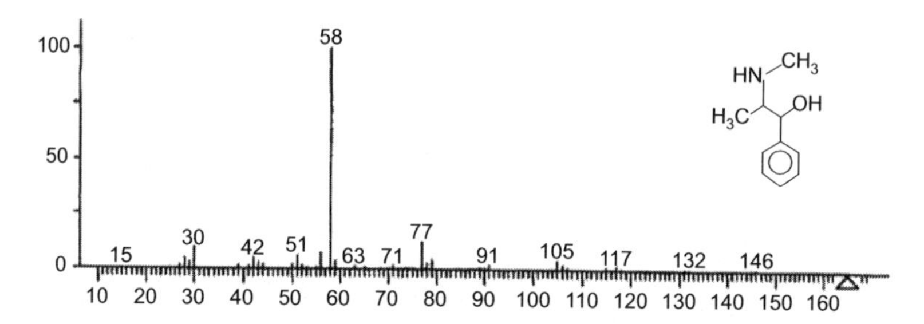

Figure 1.9 The electron ionization mass spectrum of ephedrine. *Source*: McLafferty F. W., *Interpretation of Mass Spectra.*, Third Edition. University Science Books, 1980.

of ephedrine (Figs. 1.9 and 1.10). As noted, the electron ionization spectra is devoid of a molecular ion since alpha-cleavage is the dominate fragmentation mode. This gives rise to the fragment at m/z 58 which is shown in Fig. 1.9. In the CI spectrum (Fig. 1.10), the $(M + H)^+$ ion appears in considerable abundance at m/z 166. Other important ion fragments also appear in the CI spectrum. Among these include fragments at m/z 148 $[(M + H) - H_2O]^+$ and m/z 135 $[M - (NHCH_3)]^+$. These fragment ions offer support for the presence of the –OH and –N–(CH_3) groups, respectively.

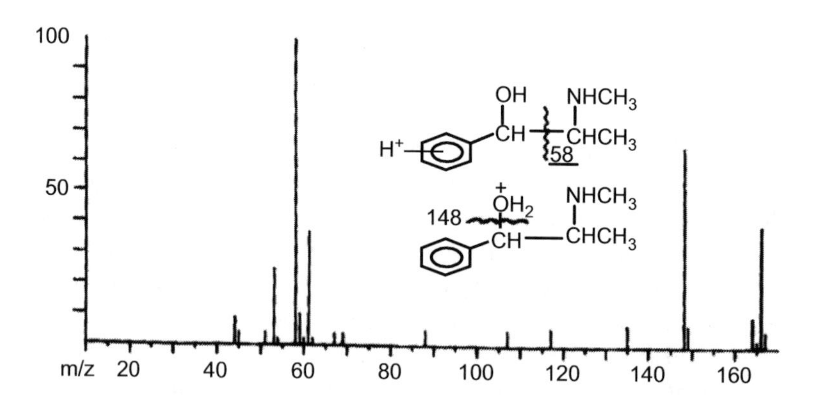

Figure 1.10 The chemical ionization (CI) mass spectrum of ephedrine using CH_5^+ as the "reagent ion." *Source*: McLafferty F. W., *Interpretation of Mass Spectra*, Third Edition. University Science Books, 1980.

1.14 High-Molecular-Weight Impurities

High mass fragments and fragments arising from impurities in the sample may be incorrectly labeled as the molecular ion. Obviously, a history of the sample along with a determination of its purity, as provided by gas chromatography or high performance liquid chromatography (HPLC), will help clarify such matters. In addition, other information such as functional group and elemental analyses could provide information relative to the tendency of the sample to lose a neutral fragment, as well as the expectation of an isotopic clusters, etc. In addition to assisting in the identification of M^+, information of this type is essential to the overall identification of the sample compound.

1.15 The Fragmentation of M^+

From a correlation of the major fragments in a mass spectrum, it may be possible to establish the identity of M^+ in an indirect fashion even though it may not be observed. For instance, suppose that the highest mass in the spectrum is at m/z 100, and there is also another mass fragment at m/z 15. This could suggest that a heavier ion is losing a methyl group, which is a preferred mode of fragmentation. If this mode of fragmentation is established, then the molecular mass could be 115. In a different example, suppose the parent molecule is known to contain oxygen, then one may suspect the loss of water and the presence of an $(M - 18)^+$ fragment. This is again, a logical expectation since alcohols and other oxygen containing molecules are known to form $(M - 18)^+$ ions. These compounds are also known to form $(M - OH)^+$ and $(M - O)^+$ fragments of relatively low intensities. The presence of these latter fragments could possibly help establish the molecular mass, even if M^+ is not present in the spectrum.

Acetates are known to rearrange, producing $(M - 60)^+$ ions by eliminating acetic acid; therefore, an ion at m/z 60 may suggest such a mode of fragmentation. If the compound is known to be an acetate, it is likely that the value of M^+ could be confirmed from such information. (See Appendix A for a list of neutral fragments that may be expelled from molecules and fragment ions as a result of electron bombardment.)

As previously indicated, information regarding the loss of one or more fragments (or neutral masses) from M^+ may be sufficient to establish the molecular mass of the sample compound without an observable molecular ion. However, molecular ion verification by fragment losses should be based upon well-established fragmentation patterns.

As mentioned in Section 1.13, the $(M + H)^+$ ion is pressure dependent, and its formation is more pronounced when the chemical ionization technique is used. However, the presence of $(M + H)^+$ ion is also facilitated by low voltage on the accelerator plate. This results in an increase in the number of bimolecular collisions by increasing the residence time of the ions within the ionization chamber. Under these circumstances, $(M + H)^+$ ions may appear under normal electron bombardment procedures. The $(M + H)^+$ ion is often more stable than M^+. If one is not careful, $(M + H)^+$ can easily be mistaken for M^+ or its ^{13}C isotopic peak. $(M + H)^+$ ions may appear alone or in the presence of M^+. If present with M^+ and if its intensity is significant, this could result in an incorrect value for the carbon number (Section 1.26) since $(M + H)^+$ and the ^{13}C isotopic peak, $(M + 1)^+$, would overlap (carbon numbers are discussed in greater detail in Section 1.26).

Compounds that are easily decomposed such as ethers, esters, amines, aminoesters, and nitriles, to name a few, may show $(M + H)^+$ ions in their spectra.

When accurately identified, the $(M + H)^+$ ion may be an ideal substitute for M^+ and will indirectly give the molecular mass of the sample compound.

1.16 Determining the Molecular Ion from the Fragments

To obtain a better appreciation of how the molecular ion may be determined from a study of the fragments, consider the spectrum of 2-methyl-2-butanol (MW 88), shown in Fig. 1.11. For explanation purposes, let us assume that the structure of the compound is unknown, but based on other analysis, it is known to contain only

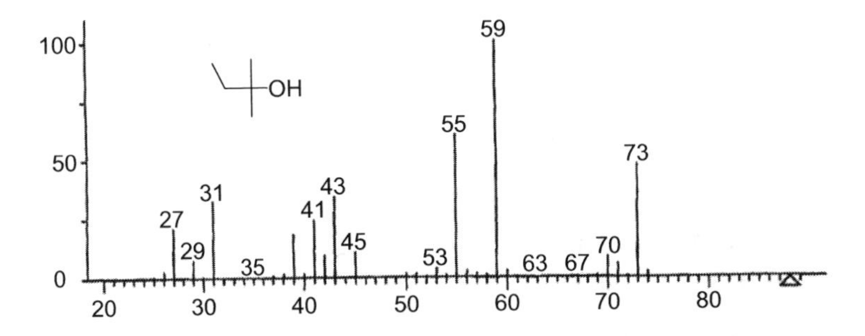

Figure 1.11 The mass spectrum of 2-methyl-2-butanol.

carbon, hydrogen, and oxygen. With the presence of one or more oxygens, we may suspect that the fragment at m/z 59 originated from the loss of water. This would lead to an M^+ value of 77 (59 + 18), which is inconsistent with the absence of nitrogen (see Section 1.7). If we suspect that the molecule has a methyl group, then the fragment at m/z 73 could represent the $(M - 15)^+$ ion. This suggests an M^+ value of 88, which is consistent with prior information. The tentative value of 88 must now be rejected or accepted on the basis of additional evidence.

The base peak at m/z 59 could represent the loss of an ethyl group. Thus far the 88 value for M^+ seems logical. If the compound is an alcohol, one would expect an $(M - 18)^+$ fragment at m/z 70, due to the loss of water. Sure enough, there is a low-intensity fragment at this mass value. In addition, the m/z 71 fragment could be the $(M - OH)^+$ ion, another characteristic of alcohols. By adding the masses of the related fragments mentioned above (73 + 15), (59 + 29) and (70 + 18), the value of 88 is tentatively accepted as the molecular weight of the sample compound. A check of alcohols having molecular weights of 88, along with a correlation of other fragments, produces 2-methyl-2-butanol as the probable structure of the compound.

$$CH_3CH_2-\underset{\displaystyle OH}{\overset{\displaystyle CH_3}{\overset{|}{\underset{|}{C}}}}-CH_3$$

2-methyl-2-butanol

Of course, the deductions above are made easy since we had prior knowledge of the compound's structure. Nevertheless, the example should serve to explain how one might proceed with the analysis of an unknown compound whose molecular ion is not observed.

1.17 Determining the Molecular Ion by Derivatization

In some instances, the absence of a molecular ion may be attributed to low volatility of the sample compound. If the structural class of the sample compound is known, the problem may be resolved by converting the compound to a more volatile derivative and obtaining its spectrum. The molecular ion of the original sample could possibly then be deduced in an indirect fashion.

Derivatization of highly polar compounds such as sulfonic acids, organic salts as well as some alcohols and amines, is especially important if the molecular ion is to be determined. An ideal derivative is one that is easy to form, preferably in one step so as to reduce the chance of contamination and extensive purification requirements.

1.18 Determining M^+ From Rates of Effusion

Using Graham's law of effusion and with knowledge of the rate of effusion of the sample from the reservoir into the ion source, it is possible to determine its molecular mass. (For a more detailed discussion of this procedure, the reader is referred to the bibliography at the end of the chapter.)

1.19 General Rearrangement

The tendency of molecular ions and other ion fragments to rearrange depends upon the potential energy of the ion itself. Obviously, the more energetic the ion, the greater its tendency to fragment or rearrange. In addition to simple fragmentation, practically all organic compounds containing two or more carbons will, when subjected to electron impact, produce rearranged ions. The formation of such ions may occur by one or several mechanistic pathways. Among these include skeletal regrouping, hydrogen, alkyl or aryl migration, as well as the loss of neutral fragments such as CO, CO_2, HCN, S, and CH_2O. As previously mentioned, some rearrangements are non-specific and defy, logical interpretation. *In general, it can be said that knowledge of established fragmentation and rearrangement patterns for a variety of classes of organic compounds is essential to the understanding of mass spectrometry. Listed in the appendix is a summary of fragmentation processes for a number of classes of organic compounds.*

1.20 Skeletal Rearrangement

Skeletal rearrangement involves the recombination of atoms to produce fragment ions that appear to have no obvious structural relationship to the original compound. Obviously, these reactions occur by unique rearrangement mechanisms which bear no resemblance to normal organic chemical processes. The formation of the relatively abundant $^{+}CH_2CH_3$ ion (m/z 29), in the spectrum of isobutane (Fig. 1.8), has already been mentioned as one example. Skeletal rearrangements are more common among organic compounds whose bonds are difficult to cleave. Among these include alkanes and polyhaloalkanes.

1.21 The McLafferty Rearrangement

The *McLafferty Rearrangement* is among the best understood rearrangements in mass spectrometry. It is common for a wide class of organic compounds adhering to the general structural characteristics shown below. Among the compounds in which the McLafferty rearrangement is expected to occur include certain aldehydes, ketones, esters, organic acids, carbonates, phosphates, sulfites, hydrazines, hydrazones, olefins, ketimines, and phenylalkyls to name a few.

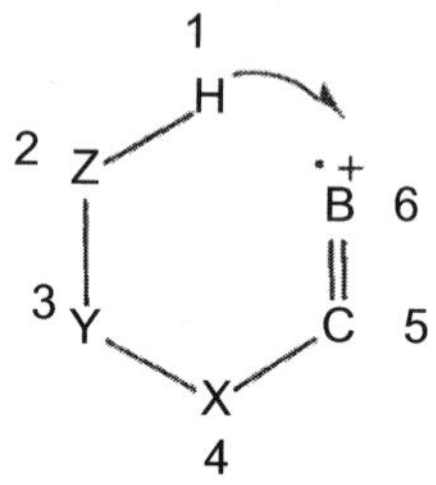

Compounds with the above general structure will undergo the McLafferty rearrangement. In the above example B, C, X, Y, and Z may be combinations of C, O, N, and S.

The McLafferty rearrangement is of an intramolecular nature and involves the transfer of a hydrogen atom from atom 2 (which is usually a carbon atom) to atom 6 (which may be a heteroatom, an olefinic carbon, or an aromatic ring). Presumably, the rearrangement occurs through a favored six-membered transition state, followed by the cleavage of bond 3–4 to eliminate a neutral fragment. An example of the McLafferty rearrangement is shown below.

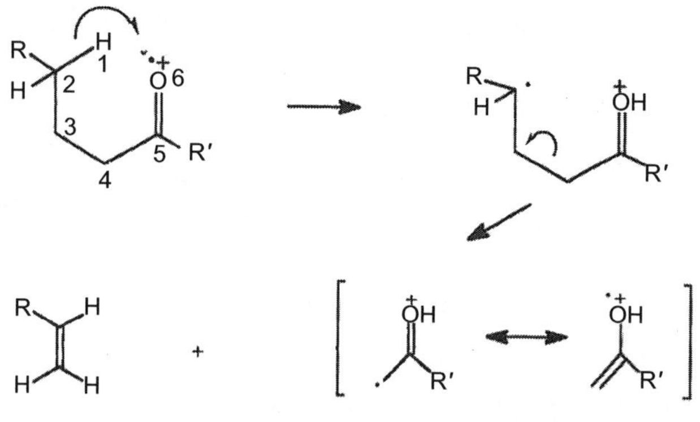

1.22 The Loss of Neutral Fragments

Rearrangements accompanied by the loss of neutral fragments are often in competition with simple cleavage reactions. Usually cleavage is the preferred mode of fragmentation; therefore, ions formed by the loss of stable neutral fragments are often in low abundances; however, this is not always the case. Ions formed by the loss of neutral fragments are usually detectable, and their presence may be predicted for certain classes of compounds. Among the stable neutral fragments most often expelled include H_2O, HF, CO, CO_2, N_2, SO_2, $CH_2{=}CO$, $CH_2{=}CH_2$, and $HC{\equiv}CH$. For instance, certain carbonates and esters are known to rearrange through a four-centered transition state, losing CO_2.

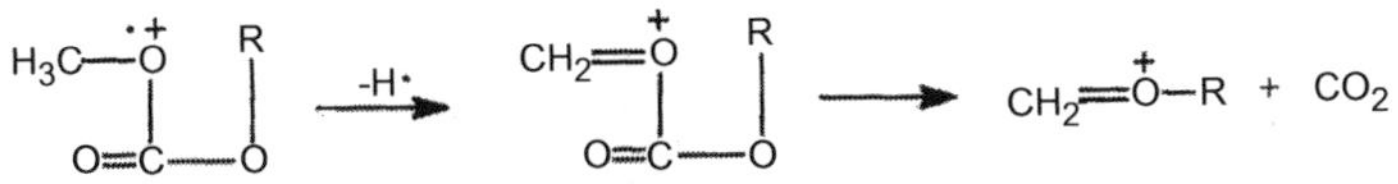

The loss of CO_2 may also occur in unsaturated esters and ethers, probably by a mechanism similar to the one shown below.

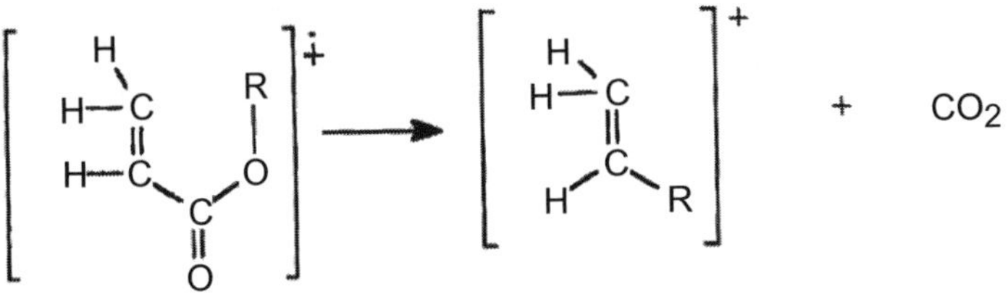

The loss of HCN and the formation of a rearranged ion, identified as $C_{11}H_{10}^{+}$, is observed in the mass spectrum of N, N-diphenylamine.

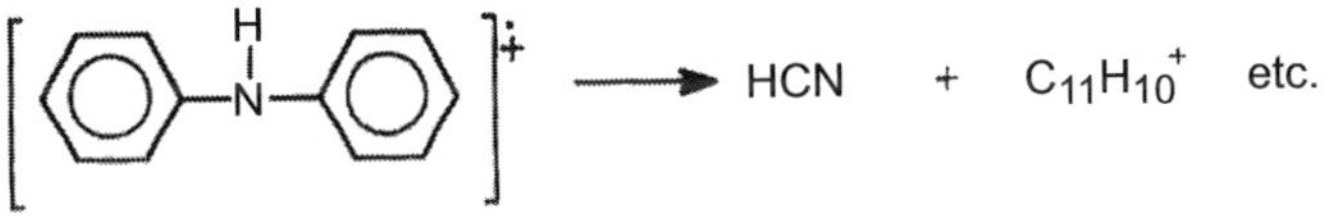

The loss of SO_2 is common among alkyl and aryl benzene sulfonamides and esters of sulfonic acids.

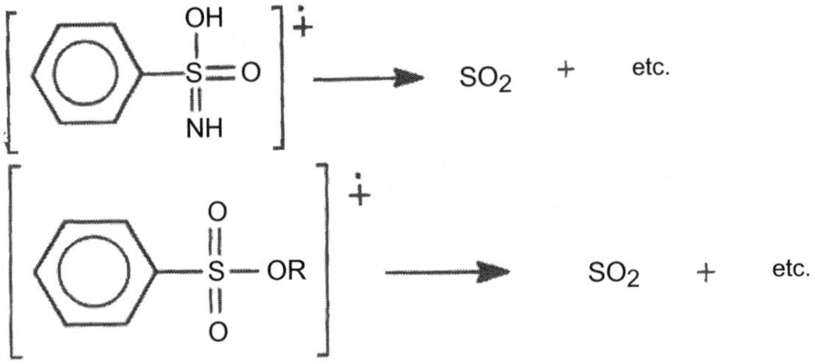

Anthraquinone loses two molecules of CO to form a $C_{12}H_8$ fragment at m/z 152.

$$\longrightarrow 2\,CO + C_{12}H_8^{+}$$

Phthalic anhydride loses CO and CO_2 to form the unstable benzyne type intermediate at m/z 76.

$$\longrightarrow CO + CO_2 + [\;]^{+}$$

Benzyne
m/z 76

Azobenzene loses N_2

N=N R

$$\longrightarrow N_2 + [C_{12}H_{10}]^{+} + \text{etc.}$$

m/z 154

The above examples represent only a few of the numerous molecular ions capable of losing a neutral fragment. A larger listing of common neutral fragments expelled from molecular ions is shown in Table 1.4.

1.23 Atomic Weight Determinations

The ability of the mass spectrometer to separate and record the masses and natural abundances of individual isotopes of some of the elements is illustrated in the expanded spectrum of mercury vapor (Fig. 1.12). By measuring the relative peak heights, the relative abundance of each of the isotopes may be determined. When the relative abundance of each isotope is multiplied by its exact mass (e.g., 197.97 rather than 198), and the results are added, the atomic

Table 1.4 Examples of some reasonable losses from molecular ions

	Odd losses	
M – 1	Hydrogen atom	M–H
M – 15	Methyl radical	M–CH_3
M – 19	Fluorine atom	M–F
M – 29	Ethyl radical	M–CH_2CH_3
M – 31	Methoxy radical	M–OCH_3
M – 35	Chlorine atom	M–Cl
M – 43	Propyl radical	M–C_3H_7
M – 45	Ethoxy radical	M–OCH_2CH_3
M – 57	Butyl radical	M–C_4H_9
M – 91	Benzyl radical	M–$CH_2C_6H_5$
M – 79	Bromine atom	M–Br
M – 127	Iodine atom	M–I
	Even losses	
M – 2	Hydrogen molecule	M–H_2
M – 18	Water	M–H_2O
M – 20	Hydrogen fluoride	M–HF
M – 28	CO or ethylene	M–CO or M–C_2H_4
M – 30	Formaldehyde	M–H_2CO
M – 32	Methanol	M–CH_3OH
M – 36	Hydrogen chloride	M–HCl
M – 44	Carbon dioxide	M–CO_2
M – 46	Nitrogen dioxide	M–NO_2
M – 60	Acetic acid	M–CH_2CO_2H
M – 90	Silanol: HO-Si $(CH_3)_3$	M–HO-Si-$(CH_3)_3$

The smallest C-containing group that can be lost is CH_3 (15) which is also, by far, the most common neutral loss peak in organic mass spectrometry. Losses >4 or <15 are not generally possible.

weight of mercury is obtained. The procedure results in a calculated value of 200.56, which corresponds favorably with the accepted value of 200.59.

1.24 The Isotopes of Carbon

Carbon has two major isotopes, ^{12}C and ^{13}C. The contributions of each of the two isotopes are usually obvious. For instance, the spectrum of methane (Fig. 1.13) shows masses based on both $^{12}C^1H_4$

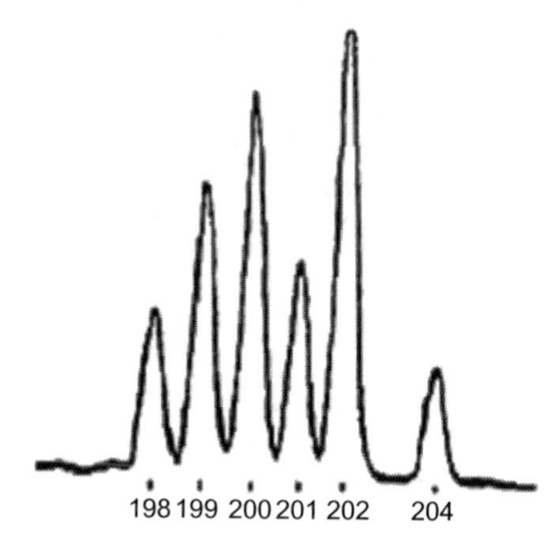

Figure 1.12 The expanded mass spectrum of mercury vapor. *Source*: Varian Associates, Palo Alto, California.

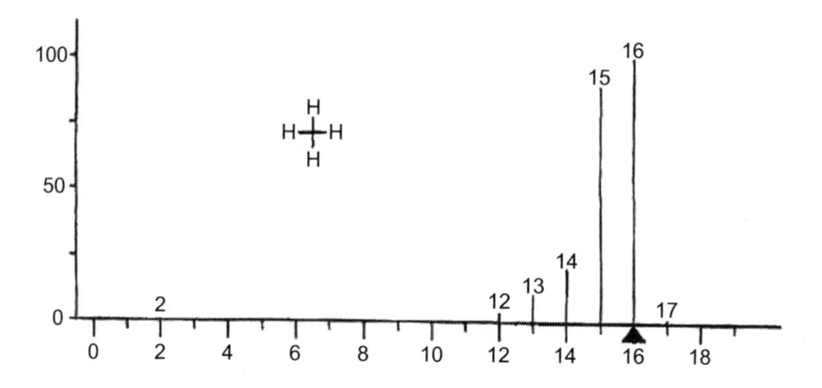

Figure 1.13 The mass spectrum of methane.

and $^{13}C^{1}H_4$. As shown in Fig. 1.13, the molecular ion of methane appears at m/z 16, a value corresponding to the more abundant ^{12}C and ^{1}H isotopes. Also observed in Fig. 1.13 is the low-intensity fragment at m/z 17. This fragment corresponds to $^{13}CH_4^{+}$, which is the $(M + 1)^+$ isotope peak. The low relative intensity of the fragment reflects the low natural abundance of the ^{13}C isotope (1.1%) and the small number of carbons in the molecule.

As previously mentioned, the intensity of the $(M + 1)^+$ peak relative to $(M)^+$ is important in approximating the number of carbons in an unknown compound. For instance, in the high mass region of the spectrum of ethane there are two peaks corresponding to M^+ and $(M + 1)^+$. The more abundant M^+ peak (at m/z 30) corresponds to molecules of ethane containing the more abundant ^{12}C and ^1H isotopes which is represented by $(^1H_6{}^{12}C^{12}C)^+$. The peak of lesser intensity at m/z 31 is the ^{13}C isotopic peak, and it represents those molecules of ethane containing both a ^{12}C and ^{13}C carbon $(^1H_6{}^{12}C^{13}C)^+$. Strictly from a theoretical viewpoint, there should be a fragment ion at m/z, 32 which represents the $(M + 2)^+$ ion. This peak represents the very small fraction of ions containing two ^{13}C atoms $(^1H_6^{13}C^{13}C)^+$. Since only a relative few molecules of ethane have two ^{13}C carbons, the intensity of the $(M + 2)^+$ peak is too small to be observed. As the number of carbons in the molecule increases, so does the relative intensities of $(M + 1)^+$ and $(M + 2)^+$. A summary of the three expected fragments and their isotopic composition is shown below:

$(^1H_6{}^{12}C^{12}C)^+$	M^+, m/z 30
$(^1H_6{}^{12}C^{13}C)^+$	$(M + 1)^+$, m/z 31
$(^1H_6{}^{13}C^{13}C)^+$	$(M + 2)^+$, m/z 32 (not observed)

It should be noted that deuterium exists in only 0.016% natural abundance and, therefore, it normally contributes very little to the intensity of $(M + 1)^+$. For molecules containing a relatively large number of hydrogen atoms, the contribution of deuterium becomes significant. If not taken into consideration, this factor could induce an error in the measured intensity of $(M + 1)^+$.

1.25 Calculating Relative Intensities

The expected relative intensity of M^+, $(M + 1)^+$, and $(N + 2)^+$, and other satellite ions may be determined from the coefficients of the expanded binomial, $(a + b)^n$, where a = natural abundance of the lighter isotope; b = natural abundance of the heavier isotope; and n = number of carbons in the molecule.

Considering ethane as the example, one may proceed to substitute the proper values of a, b, and n into the expanded binomial to obtain the following results.

$$
\begin{aligned}
(a+b)^2 &= a^2 &&+\ 2ab &&+\ b^2 \\
&= (98.98)^2 &&+\ 2(98.98)(1.1) &&+\ (1.1)^2 \\
&= 9779 &&+\ 219.5 &&+\ 1.23
\end{aligned}
$$

or

$$
= 100 \quad + 2.2 \quad + 0.013
$$

Therefore,

$$
M^+ = 100 \qquad (M+1)^+ = 2.2 \qquad (M+2)^+ = 0.013
$$

As we can see, for ethane the calculated intensity of $(M + 2)^+$ is too small to be observed, but as indicated, the ratio of the intensity of both $(M + 1)^+$ and $(M + 2)^+$ to M^+ will increase as the number of carbons increase.

Realistically, the calculation of relative peak intensities of known compounds is of little use in the analysis of mass spectral data. The procedure is presented only to give the reader an appreciation of the more important reverse operation which involves approximating the number of carbons atoms in an unknown compound from the relative intensities of M^+ and $(M + 1)^+$. This matter is discussed in more detail in Section 1.26

1.26 The Experimental Determination of the Carbon Number

As an approximation, each carbon atom in a molecule contributes about 1.08 to the intensity of the $(M + 1)^+$ ion (relative to M^+). It is, therefore, possible to approximate the number of carbon atoms in an unknown compound (or fragment) by a careful measurement of the relative heights of M^+ and $(M + 1)^+$. However, it should be emphasized that the resulting calculation is only a rough approximation. As an application of the method, consider the spectrum of n-butyrophenone shown in Fig. 1.14.

In the expanded portion of the spectrum, M^+ and $(M + 1)^+$ are shown at m/z 148 and 149, respectively. If the abundance of M^+ and $(M + 1)^+$ are adjusted such that M^+ is 100%, a value of 11% is obtained for $(M + 1)^+$. By dividing 11 by 1.08, the approximate number of carbons atoms in the molecule is calculated to be 10, which is consistent with the formula of n-butyrophenone.

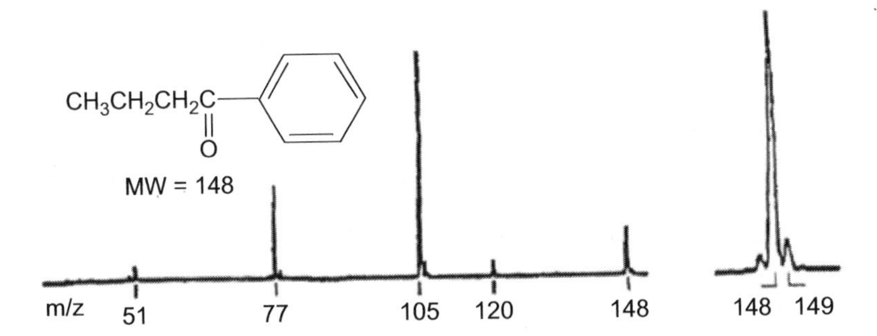

Figure 1.14 The mass spectrum of n-butyrophenone, showing the expanded M^+ and $(M+1)^+$ region.

In the case of benzothiazole (Fig. 1.5), M^+ was also the base peak; however, as we have seen, this is not always the case. For instance, consider the following partial mass and abundance data for acetophenone (C_8H_8O). Acetophenone has a molecular weight of 120, and its mass spectrum is shown in Fig. 1.3.

m/z	**Abundance**	**Adjusted values**
105 (base peak)	999	
120 (M^+)	333	100
121 $(M+1)^+$	29	8.7
122 $(M+2)^+$	1	0.3

In this example, M^+ is adjusted to 100% abundance, followed by an adjustment of $(M + 1)^+$ and $(M + 2)^+$. This is accomplished by dividing the abundance of M^+ into itself and into the abundance of $(M + 1)^+$ and multiplying the results by 100. As noted above, when the adjustment is made, the value for $(M + 1)^+$, relative to M^+, is 8.7. By dividing 8.7 by 1.08, 8 carbon atoms are approximated for the molecule. The procedure for calculating the carbon number may be expressed by the following equation:

$$\frac{\dfrac{\text{Relative abundance of }(M+1)^+}{\text{Relative abundance of }M^+}\times 100}{1.08} = \text{The carbon number}$$

The fact that the $(M+2)^+$ ion is rather small suggests the absence of bromine, chlorine, sulfur or silicone. These elements have relatively large $(M+2)^+$ isotopes whose presence would result in a value for $(M+2)^+$, much larger than the value listed above.

Since oxygen, hydrogen, and nitrogen all have $(M+1)$ isotopes, it is expected that these elements, like ^{13}C, will increase the value of $(M+1)^+$ relative to M^+. However, the relative abundances of these elements are quite small (see the appendix), and they do not contribute appreciably to the intensity of the $(M+1)^+$ peak. However, as the number of hydrogens in the sample molecule increases, the contribution of deuterium becomes progressively more significant. For instance, deuterium exists in only about 0.016% abundance; therefore, each hydrogen adds approximately 0.016% to the intensity of $(M+1)^+$, relative to M^+. As an example, consider the spectrum of hexadecane shown in Fig. 3.1. In this compound the 16 carbons should contribute 17.28%, (16×1.08) to the intensity of $(M+1)^+$ relative to M^+, while the contribution of deuterium should only be 0.544%, (34×0.016). Consequently, the calculated percentage of the intensity of $(M+1)^+$ relative to M^+ should approximate 17.82%.

Regarding ^{15}N and ^{17}O, their natural abundances are 0.38 and 0.04, respectively. Hence, for a compound with the formula $C_5H_{13}O_2N$, the calculated relative percentage abundance of $(M+1)^+$ should be $(5 \times 1.08$ for ^{13}C or $5.4) + (13 \times 0.016$ for ^2H or $0.208) + (2 \times 0.04$ or 0.08 for $^{17}O) + (1 \times 0.38$ or 0.38 for $^{15}N) = 6.06$. Taking into consideration the low probability of sample molecules containing two of the minor isotopes of C, H, and O, one can obtain similar calculations for the $(M+2)^+$ isotopic peak for a wide variety of empirical formulas containing these elements. Such information, as we shall see in Section 1.30, is important in selecting the appropriate molecular formula from several different formulas having the same molecular mass.

The procedure for calculating the number of carbon atoms in a sample molecule represents an approximation which progressively loses some of its accuracy as the number of carbons in the molecule increases beyond 15. This loss of accuracy, as indicated, may be attributed mainly to the increasing contribution of deuterium to the intensity of $(M+1)^+$ as the number of hydrogens in the molecule increases. Other possible factors affecting the accuracy include the presence of $(M + H)^+$ ions and/or the presence of nitrogen in the molecule as well as impurities whose fragments may be superimposed upon M^+ or $(M + 1)^+$. The presence of either of these aforementioned factors, if not taken into consideration, could result in an error in the measured intensities and in the calculated carbon number.

1.27 Compounds Containing Bromine and/or Chlorine

In addition to carbon, the number and kinds of certain other elements present in sample compounds may be determined provided that the elements have sufficient natural abundances. An observation of Tables 1.1 and 1.5 indicates that chlorine and bromine, and to some extent sulfur and silicon, are elements which are suited for these determinations. For instance, bromine has two

Table 1.5 The percentage natural abundances of some frequently occurring isotopes

Element	Natural abundance (%)
^{12}C	98.89
^{13}C	1.08
^{1}H	99.985
^{2}H	0.015
^{14}N	99.64
^{15}N	0.36
^{16}O	99.76
^{17}O	0.04
^{18}O	0.20
^{35}Cl	75.77
^{37}Cl	24.23
^{79}Br	50.69
^{81}Br	49.31
^{32}S	95.0
^{33}S	0.76
^{34}S	4.22
^{36}S	0.02

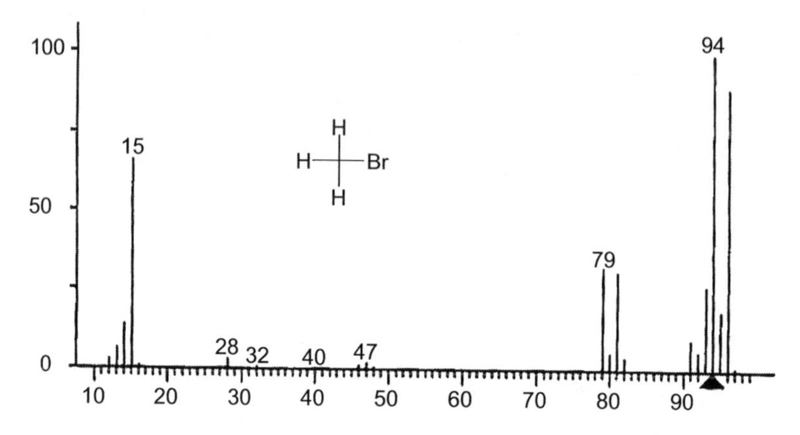

Figure 1.15 The spectrum of methyl bromide showing the M^+ and $(M+2)^+$ peaks at m/z 94 and 96, respectively.

major isotopes which exist in almost equal abundances (50.69% and 49.31%) with differences of two mass units. Therefore, the spectrum of compounds containing one bromine atom would be expected to show an isotopic cluster consisting of M^+ and $(M + 2)^+$ of about equal intensities, as illustrated in the spectrum of methyl bromide (Fig. 1.15).

As the number of bromine atoms in the sample compound increases, so does the complexity of the isotopic cluster in the high mass region. For instance, compounds containing two bromine atoms should reveal, among other fragments, a duster consisting of three isotopic fragments, located at M^+, $(M + 2)^+$ and $(M + 4)^+$. The spectrum of 1,2-dibromoethane, for example, should reveal ions having the following general isotopic composition and m/z values:

Isotopic composition	Ion fragment	m/z value
$H_4C_2{}^{79}Br_2$	M^+	186
$H_4C_2{}^{79}Br^{81}Br$	$(M+2)^+$	188
$H_4C_2{}^{81}Br^{81}Br$	$(M+4)^+$	190

In addition to the three isotopic ions of dibromoethane, one would also expect to observe a low-intensity ^{13}C isotopic ion at $(M+1)^+$.

As illustrated in section 1.25, the relative intensity and number of fragment ions making up an isotopic cluster may be predicted from an expansion of $(a + b)^n$. When the procedure is applied to dibromomethane (where a and b are the relative natural abundances of ^{79}Br and ^{81}Br, respectively, and n = the number of bromine atoms in the molecule), the following relationship is obtained:

$$\begin{array}{lllll} a^2 & + & 2ab & + & b^2 \\ 1^2 & + & 2(1)(1) & + & 1^2 \\ M^+ & + & (M+2)^+ & + & (M+4)^+ \end{array}$$

Therefore,

$$M^+ = 1; \quad (M+2)^+ = 2; \quad (M+4)^+ = 1$$

Note: Because the isotopes of bromine exist in about equal abundances, we have simplified matters by substituting in the expanded binomial, a value of 1 for the abundance of each of bromine's isotopes.

From the expansion of the binomial, we can see that dibromoethane, as well as any compound containing two bromine atoms, has an isotopic cluster made up of three ions M^+, $(M + 2)^+$, and $(M + 4)^+$, with intensity ratios of approximately 1 : 2 : 1. As indicated by Fig. 1.16, calculations of this nature have been made for compounds containing from one to six bromine atoms. By comparing these patterns with isotopic clusters found in actual spectra, it is often possible to determine if the compound contains bromine and the number of bromine atoms present. Although the patterns in Fig. 1.16 may vary slightly from actual spectra, the similarity is often close enough to make an accurate bromine number determination.

Compounds containing chlorine as the only halogen may be treated in the same fashion as described for bromine. However, one would expect the isotope cluster to be different from those of bromine, since the natural abundances of the two isotopes are

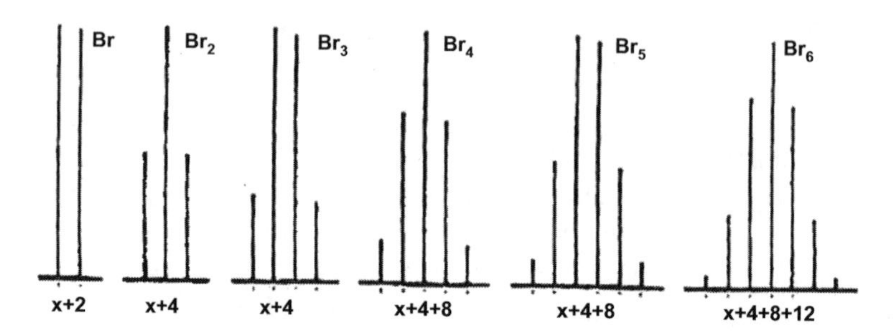

Figure 1.16 The group patterns characteristic of molecules containing one to six bromine atoms. *Note*: The peaks are at intervals of 2 m/z (X, X+2, X+4 ...). As an example, an ion with 3 bromine atoms will have an isotopic cluster with peaks at X, X+2, X+4, and, X+6. These peaks will be in the ratio of approximately 34 : 100 : 98 : 32. Also see Table A.6 in the Appendix.

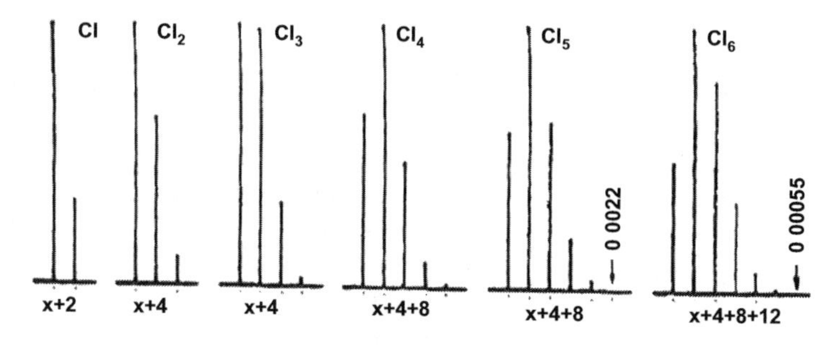

Figure 1.17 The group patterns characteristic of molecules containing one to six chlorine atoms. *Note*: The peaks are at intervals of 2 m/z (X, X+2, X+4, ...).

different. Figure 1.17 shows the expected isotopic cluster patterns for compounds containing one to six chlorine atoms.

For compounds containing both chlorine and bromine, the number of peaks making up the isotopic cluster and their intensity ratios may be calculated by expanding a slightly different binomial, given by

$$(a + b)^n (c + d)^m,$$

where a = the relative abundance of the lighter isotope of bromine; b = the relative abundance of the heavier isotope of bromine; c = the relative abundance of the lighter isotope of chlorine; d = the relative abundance of the heavier isotope of bromine; n = the number of bromine atoms in the molecule; and m = the number of chlorine atoms in the molecule.

Consider the following calculations for 1-chloro-2-bromoethane:

$$(a + b)(c + d) = ac + ad + bc + bd,$$

where ac refers to $H_4C_2{}^{79}Br^{35}Cl = M^+$; ad refers to $H_4C_2{}^{79}Br^{37}Cl = (M + 2)$; bc refers to $H_4C_2{}^{81}Br^{35}Cl = (M + 2)$; and bd refers to $H_4C_2{}^{81}Br^{37}Cl = (M + 4)$.

Using a 1 : 1 relationship for the natural abundance for the isotopes of bromine and a 3 : 1 ratio for those of chlorine and by substituting the correct values in the expanded binomial, the following results are obtained:

$$ac + ad + bc + bd = (1 \times 3) + (1 \times 1) + (1 \times 3) + (1 \times 1)$$

$$M^+ = 3$$

$$(M + 2)^+ = 1$$

$$(M + 2)^+ = 3$$

$$(M + 4)^+ = 1$$

Both ad and bc contribute to the intensity of $(M + 2)^+$; consequently these values should be added, giving an overall intensity ratio of $M^+ = 3$, $(M + 2)^+ = 4$, and $(M + 4)^+ = 1$. The characteristic appearance of the isotope clusters for compounds containing combinations of chlorine and bromine atoms are shown in Fig. 1.18 and Appendix A.

In summary, the mass spectra of compounds containing bromine and/or chlorine reveal characteristic and conspicuous isotopic

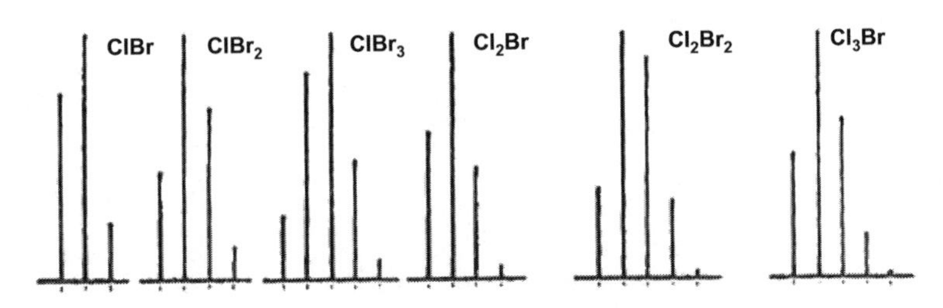

Figure 1.18 The group patterns characteristic of molecules containing combinations of bromine and chlorine atoms. The peaks are at intervals of 2 m/z (X, X+2, X+4, . . .).

cluster patterns, making it possible to determine the number of bromines and/or chlorines atoms in the unknown sample.

While discussions of isotope clusters have been more or less limited to molecular ions and their isotope satellites, it should be emphasized that similar consideration may also be applied to other halogen containing fragment ions of lesser mass.

1.28 Compounds Containing Sulfur

Sulfur has two major isotopes: ^{32}S (95% abundance) and ^{34}S (4.2% abundance). Compounds containing one sulfur atom should show an $(M+2)^+$ fragment whose intensity is about 4% of M^+. For molecules containing two sulfur atoms, the intensity of $(M+2)^+$ should be about 8% of M^+. If in addition to sulfur the compound contains one or more bromine or chlorine atoms, the isotopic cluster becomes more complex.

In the expanded high mass region of the spectrum of benzothiazole (Fig. 1.19), the $(M+1)^+$ ion at *m*/*z* 136 is attributed to $(^{13}CC_6H_5NS)^+$. This ion fragment represents the contribution of ^{13}C. A careful measurement of the intensity of this fragment shows it to be about 8% of the intensity of M^+. Dividing 8 by 1.08 (the contribution of each carbon to the relative intensity of the $(M+1)^+$ ion), an approximate value of 7 is obtained for the number of carbons in the molecule. In addition, the value of $(M+2)^+$ is

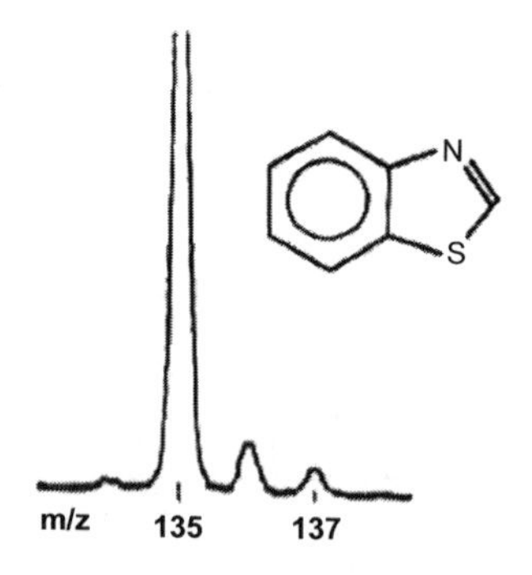

Figure 1.19 The expanded high mass region of the spectrum of benzothiazole.

about 5% of the intensity of M^+, suggesting the presence of one sulfur. It should also be noted that M^+ has an odd numerical value which means that the molecule has an odd number of nitrogen atoms. Assuming that the compound has one nitrogen (rather than three or five), we may proceed to write a partial formula of C_7NS for the compound which adds up to a mass of 130. By adding five hydrogens, a completed formula of C_7H_5NS is tentatively obtained for the compound.

The above example demonstrates the importance of M^+, its isotopic peaks, and its relative intensities in structural determinations. However, not all compounds are as easily determined. The complete spectrum of benzothiazole is shown in Fig. 1.5.

Fluorine, iodine, and phosphorus are monoisotopic elements and make no contribution to the intensity of $(M + 1)^+$ and $(M + 2)^+$. The presence of these elements may sometimes be deduced by (a) observing ions at m/z 19 (fluorine), m/z 127 (iodine), and m/z 31 (phosphorus) or (b) by a relatively small $(M + 1)^+$ ion and (c) by studying the ion fragments.

1.29 Beynon's Table

A set of tables published by J. H. Beynon in 1960 and expanded in 1963 has proven to be useful in determining the empirical formulas of unknown compounds from low-resolution mass spectra data, or more often in reducing the number of possible choices. The tables consist of calculated values of $[(M + 1)^+/M^+] \times 100$ and $[(M + 2)^+/M^+] \times 100$. The empirical formulas listed in Beynon's tables contain combinations of C, H, O, and N, along with masses between 12 and 500 mass units. A portion of Beynon's table for mass 80 is shown in Table 1.6.

The successful application of the method requires that the mass spectrum of the sample compound provides at least an M^+ and $(M + 1)^+$ ion fragment of measurable intensity. Secondly, there must be no interfering factors that would incorrectly affect the intensity of M^+ and $(M + 1)^+$, such as (a) poor resolution, (b) the presence of impurity ions having m/z values equal to M^+ or $(M + 1)^+$, (c) the presence of $(M + H)^+$ ions, and (d) intense $(M - H)^+$ whose C^{13}

Table 1.6 A portion of Beynon's table for mass 80

		Intensities	
		$[(M + 1)^+/M^+] \times 100$	$[(M + 2)^+/M^+] \times 100$
1.	CH_4O_4	1.30	0.80
2.	$C_3H_2N_3$	4.42	0.07
3.	C_4H_2NO	4.78	0.29
4.	$C_4H_4N_2$	5.15	0.11
5.	C_5H_4O	5.51	0.32
6.	C_5H_6N	5.88	0.14
7.	C_6H_8	6.61	0.18

fragment would overlap with M^+. Even in the absence of interfering factors, the actual relative intensities are usually slightly higher than those found in the tables. As an illustration of the use of the Beynon's table, consider an unknown compound whose molecular mass is 80 and whose mass spectrum reveals the following information:

$$\frac{(M+1)^+}{M^+} \times 100 = 5.18\%$$

$$\frac{(M+2)^+}{M^+} \times 100 = 0.13\%$$

Previously, we mentioned that the presence of chlorine, bromine or sulfur in a compound may be determined by the relatively large $(M + 2)^+/M^+$ ratio. Since this value in the example above is relatively small, it may be concluded that these elements are not present in the compound. By dividing 5.18 by 1.08 we can conclude that the compound contains between 4 and 5 carbons. Of the molecular formulas shown in Table 1.6, formulas 2, 3, and 6 may be immediately eliminated on the basis of the nitrogen rule. It is also reasonable to eliminate formulas 1 and 7 since their relative intensities are out of line with the above data. This leaves only formulas 4 and 5 as possible formulas. Out of these two, formula $C_4H_4N_2$ fits the data best and is tentatively selected as the empirical formula of the compound. Additional support for the elimination of formula C_5H_4O may be obtained by considering other fragments in the spectrum. For instance, oxygen compounds, as previously mentioned, usually produce characteristic fragments at $(M - 18)^+$ due to the loss of water. If the compound is an alcohol, one would expect a prominent $CH_2{=}OH^+$ at m/z 31. The absence of these fragments suggests the absence of oxygen; however, care should be exercised in arriving at a definite conclusion.

The intensity of M^+ also affords a clue to the structure of the compound. As an example, if the formula $C_4H_4N_2$ is cyclic and if branching is absent, then it should produce a more intense molecular ion than a branched, non-cyclic compound of the same formula.

In cases where the molecular ion is suspected of containing atoms other than C, H, N, and O, an adjustment must first be made to both the mass of M^+ and its isotopic ratios, before Beynon's table can be used. To illustrate the application, consider the following information obtained from the spectrum of an unknown silicon containing compound of molecular mass 188:

$$\frac{(M+1)^+}{M^+} \times 100 = 16.50\%$$

$$\frac{(M+2)^+}{M^+} \times 100 = 0.82\%$$

We must now subtract 28 (which is the mass of the lighter and more abundant silicon isotope) from 188 (the mass of M^+). This leaves a value of 160 as the adjusted molecular mass of the compound. Since the heavier silicon atom ^{29}Si exists in approximately 4.67% natural abundance, it contributes approximately 0.7 units to the value of 100 $\times$ $[(M+1)^+/M^+]$. Assuming only one silicon atom exists, we subtract 4.7 from 16.50 to obtain the following:

$$\frac{(M+1)^+}{M^+} \times 100 = 11.80\%$$

$$\frac{(M+2)^+}{M^+} \times 100 = 0.82\%$$

Beynon's table may now be consulted for mass 160, and one may now proceed in the normal fashion. Once the probable empirical formula has been determined, silicon must be added in order to complete the elemental composition of the compound. Compounds containing F, I, or P are monoisotopic and require no adjustment to the relative intensities; however, the atomic masses of these atoms should be subtracted from the mass of the molecular ion before consulting the tables.

1.30 Some Techniques, Innovations, and Applications in Mass Spectrometry

1.30.1 Thermogravimetric Analysis–Mass Spectrometry (TG/MS) and Thermogravimetric Analysis–Fourier Transform Infrared (TG/FTIR)

As a background for the discussion of combined thermogravimetric analysis–mass spectrometry or TG/MS, it may be appropriate to briefly describe the common thermal methods. Thermal analysis itself may be defined, generally, as an investigation of the effects of temperature changes on a substance. Thermoanalytical techniques are valuable in understanding the thermal properties of a variety of materials. The technique has found use in many areas such as pyrotechnical studies, polymer studies, explosive research, soil evaluation, environmental and mineralogical studies, and metallurgical and material testing. Among the most common thermal techniques are thermogravimetry or thermogravimetric analysis (TG), differential thermal analysis (DTA), derivative thermogravimetric analysis (DTG), and differential scanning calorimetry (DSC).

TG measures the weight loss of the sample as it is heated under controlled conditions of temperature and pressure (Fig. 1.20). These studies may be conducted under normal, increased, or reduced pressures and in different gas atmospheres. Valuable information relating to the thermal stability and composition of the sample material may be obtained from such studies.

DTA compares the temperature changes in the sample verses an inert material as both materials are heated (or cooled) at a uniform rate. Temperature changes in a sample may be either endothermic or exothermic, depending upon the nature of the physical or chemical changes involved. Typical among these changes include melting, boiling, changes in crystalline form, oxidation, reduction, decomposition, phase transitions, and dehydration. Endothermic effects usually accompany phase transitions, dehydrations, reductions, and certain decomposition reactions. Among the common exothermic processes are oxidation and crystallization.

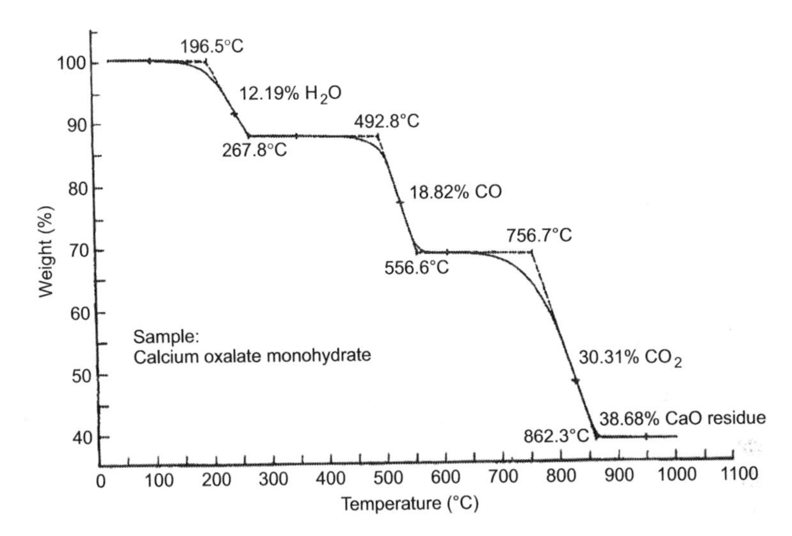

Figure 1.20 A TG curve for the thermal decomposition of calcium oxalate monohydrate.

Derivative thermogravimetric analysis (DTG) provides the first derivative of the mass curve (dm/dt). In other words, the DTG curve is the TG curve expressed in a different visual form. Information obtained from the DTG curve may be used to extract kinetic information relating to the thermal decomposition of the sample material. It is also useful in resolving subtle thermal phenomena that may not be obvious from an examination of the TG curve (compare the curves in Fig. 1.21).

Differential scanning calorimetry (DSC) measures heat flow in or out of the sample (as opposed to temperature differences between the sample and an inert reference, as in DTA). In DSC, the sample is heated or cooled at a linear rate, while the temperature of the sample and reference is maintained isothermally, with respect to each other by the application of electrical energy. The amount of heat energy necessary to maintain the sample isothermally is recorded on the DSC curve. The resulting curve (Fig. 1.22) is similar in appearance to the DTA curve, but it measures the heat flow rate of the sample with respect to temperature (dH/dt). The area under both the DTA and DSC curve is proportional to the change in enthalpy (ΔH).

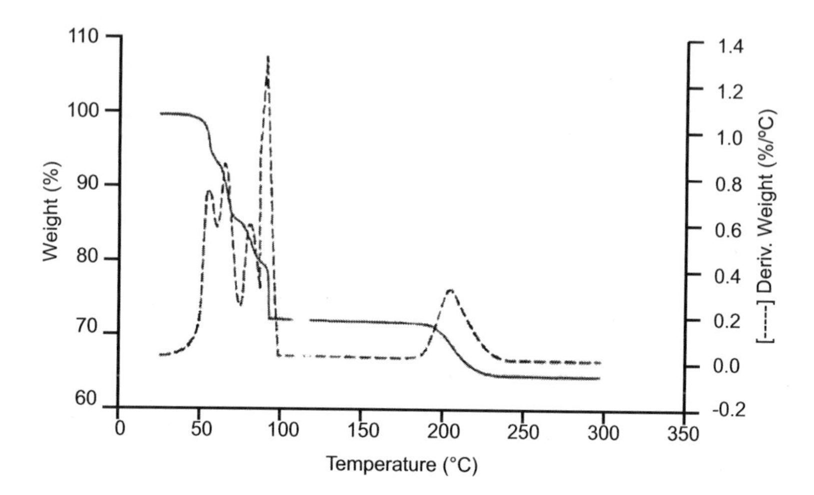

Figure 1.21 A comparison of the high-resolution TG curve (solid line) and DTG curve (dashed line) of cupric sulfate pentahydrate ($CuSO_4 \cdot 5H_2O$). *Source*: The *TA Hotline*, Volume 2, 1991, TA Instrument, Inc., New Castle, DE.

Despite the valuable information obtained from the briefly described thermal methods, none of them can provide information on the structural identity of decomposition products. Yet this information is vital to the complete characterization of material behavior under thermal conditions. By coupling thermal systems with mass spectrometers and Fourier Transform infrared spectrometers, this limitation is removed. The combined system provides more information than the systems operating separately. With the aid of sophisticated MS, FTIR and thermal analysis software, commercial MS/TG and FTIR/TG systems can provide an abundance of information relating to the thermal decompositions of materials. Some of these techniques are discussed below.

Total Evolved Gas Thermogram (FTIR): This technique provides spectral information on all evolved gases as a function of time. Consider, for example, the total FTIR gas thermogram obtained from the thermal decomposition of vinyl acetate (13%) vinyl chloride copolymer (Fig. 1.23a).

Specific Gas Profile (FTIR): In this profile a specific frequency region is scanned over time in an effort to monitor a particular gas. This

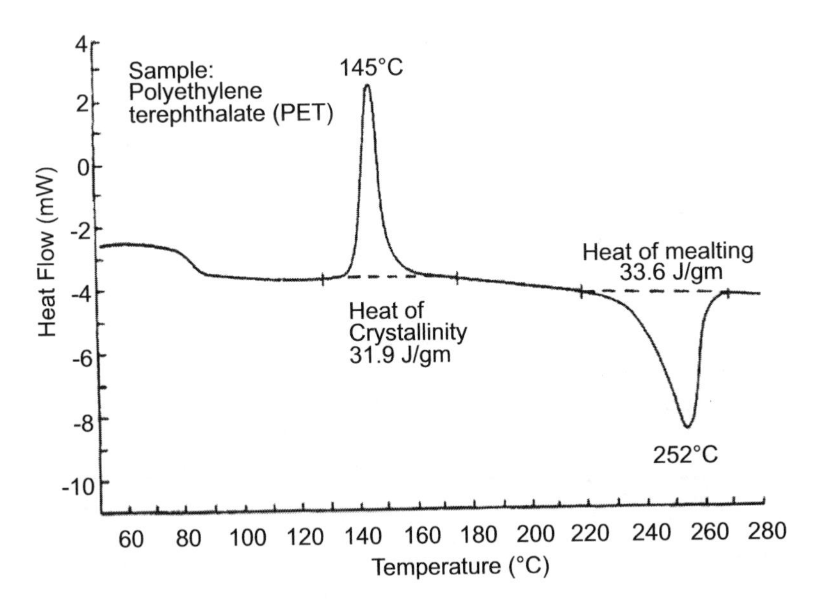

Figure 1.22 A typical DSC curve.

profile is useful for studying the evolution of specific gases. For instance, the evolution of HBr over a specific time frame may be monitored by obtaining the gas profile characteristic of the HBr stretching frequency.

***Specific Frequency Thermogram (or Functional Group Profile)*:** Here, the absorbance within a small frequency region is monitored as a function of time or temperature. This evolved gas profile is useful in monitoring the evolution of gases with common functional groups. Consider, as an example (Fig. 1.23) which shows the evolution of acetic acid and carbon dioxide, two decomposition products resulting from the thermal decomposition of the polymer vinyl acetate/polyvinyl chloride (PVA/PVC) co-polymer.

1.30.2 TG/MS Systems

Descriptions of TG/MS systems for investigating the thermal decomposition of nonmetallic materials have been reported as far back as 1968. For instance, in 1968 Zitomer reported the

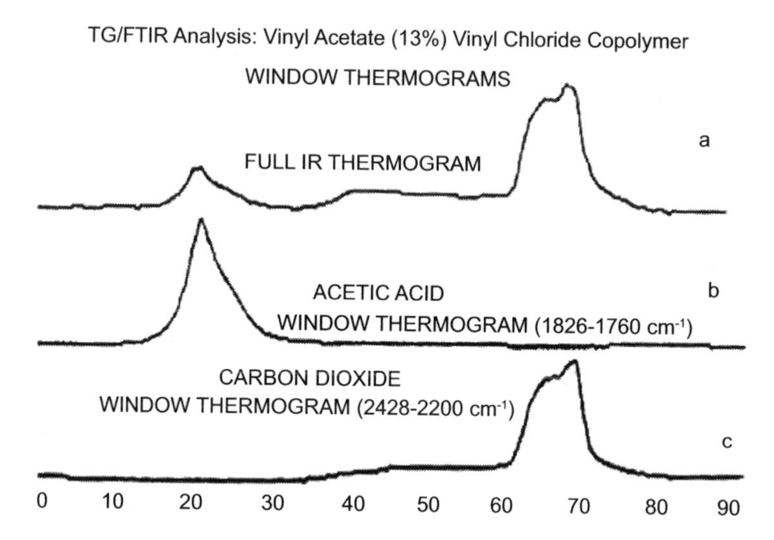

Figure 1.23 Examples of IR thermograms of vinyl acetate (13%)–poly vinyl chloride (PVA/PVC) copolymer showing the total gas thermogram (a) and the specific frequency thermograms of acetic acid (b) and carbon dioxide (c), products of the thermal decomposition. *Source*: Cassel B. and G. McClure, *Am. Lab.*, January, 1989.

first combined TG/MS system which involved the coupling of a time-of-flight mass spectrometer with a Dupont thermoanalyzer.[1] The system was used to investigate the thermal decomposition of polymethylene sulfide and maleic hydrazide methyl vinyl ether co-polymer. In 1969, Wilson and Hamaker reported the coupling of a thermoanalyzer with a quadrupole mass spectrometer.[2] Their system was used to investigate the thermal decomposition of polymethyl methacrylate, polystyrene, and polyoxymethylene. In 1972, Gibson and Johnson described the coupling of a Mettler thermoanalyzer and a quadrupole mass spectrometer.[3] This coupled system was more elaborate than the previously described systems in that it could investigate thermal decompositions under vacuum or under high pressure to one atmosphere. Chang and Mead described a TG/GC/MS system which was used in the study of polymer degradation.[4] Mettler Instruments in a company publication,[5] Langer,[6] and Wiedemann[7] all reported a coupling system for investigating TG decompositions under high vacuum. In these systems, the mass analyzer was placed directly into the vacuum chamber of the thermobalance. Wendlandt and others modified the pyrolysis chamber of an evolved gas detection/ mass spectrometer (EGD/MS) system so that DTA measurements could be made simultaneously with other thermal measurements.[8] Other DTA/EGD/MS systems have been described by Redfern[9] and Gaulin.[10] In 1974 Dunner and Eppler, in a *Balzers High Vacuum Report*, summarized the different coupling techniques between quadrupole mass spectrometers and thermoanalyzers.[11] At present, commercial TG/MS systems are available from at least 10 manufacturers, and the technique has been recognized as crucial in understanding the thermal behavior of a variety of materials.

1.30.3 The TG/MS Interface

For evolved gas studies under atmospheric pressure, a two-stage pressure reduction gas inlet system must be inserted between the sample and the chamber of the mass spectrometer. This arrangement is necessary since the evolved gases must enter the chamber of the mass spectrometer without significantly reducing the pressure inside the chamber. To obtain accurate evolved gas data, the gas inlet system must meet the following requirements:

- The ratio of the gaseous components must not change as they enter the chamber of the mass spectrometer.
- There must be a short response time between gas evolution and the time the volatile components enter the spectrometer.
- There must be little or no condensation of major gaseous components as they travel to the vacuum chamber of the mass spectrometer.

To satisfy the requirements for mass spectral studies of decomposition gases under atmospheric conditions, one end of a heated inert capillary tube is placed in the furnace of the thermogravimetric analyzer, near the sample and the other end attached to a special gas inlet valve. The heated capillary should have a small inner diameter, ranging between 0.15 and 0.30 mm. When thermal decomposition of the sample is conducted, the volatile components are swept out

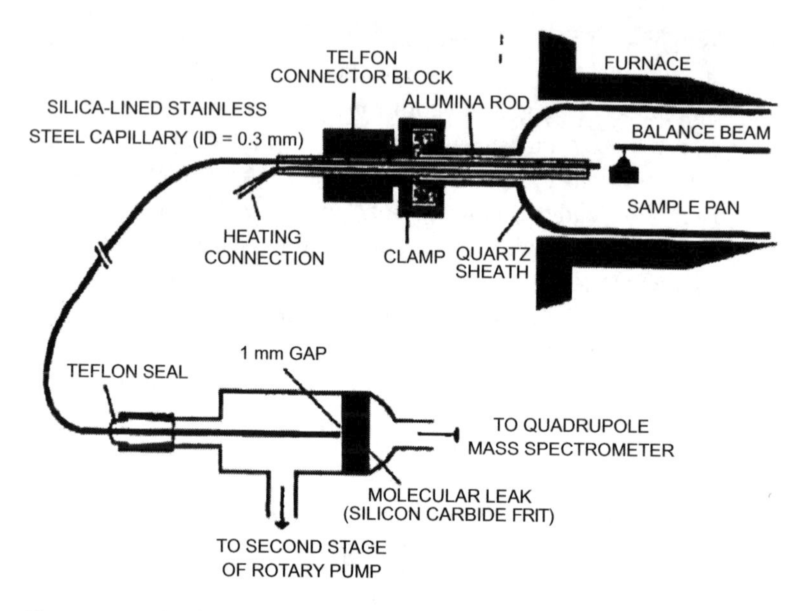

Figure 1.24 A schematic of a TG/MS interface. *Source*: Charsley E. L., S. B. Warrington, G. K. Jones, and A. R. McGhie, *Am. Lab.*, January 1990.

of the furnace by either a stream of dry air or an inert gas. The decomposition gases are then pulled through the heated inlet tube into a special inlet valve by a rotary pump attached to the inlet valve. Once the gases enter the special inlet valve, they are leaked into the chamber of the mass spectrometer through a special designed frit. The capillary inlet serves as the first pressure reduction stage and the frit as the second. The special inlet valve allows the evolved gases to enter the chamber of the mass spectrometer without significantly changing the pressure within the chamber or the ratio of gaseous components. A schematic and photograph of a TG/MS system are shown in Figs. 1.24 and 1.25, respectively.

1.30.4 TG/MS Data Collecting and Processing Software

For modern TG/MS systems, data from the mass filter of the quadrupole spectrometer is rapidly acquired and digitally transferred to the computer for processing in real-time, or the data may be stored for post-run processing. Modern TG/MS systems are

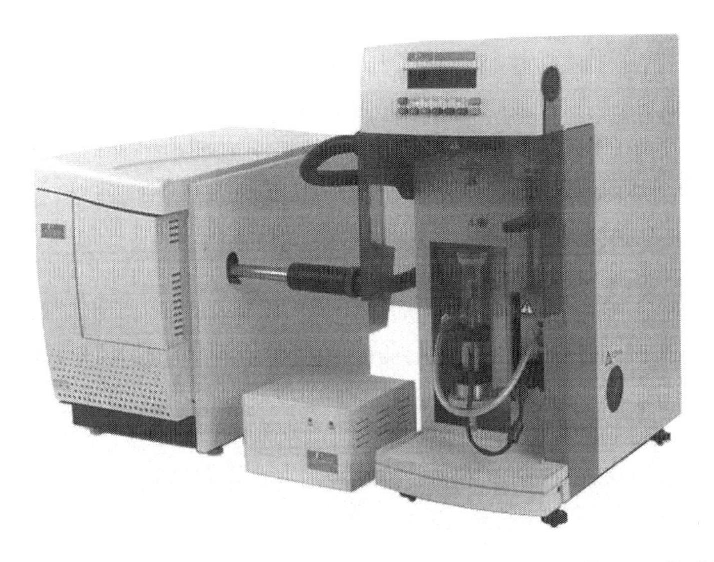

Figure 1.25 Photograph of a combined TG/MS system. *Source*: Perkin-Elmer Corporation, Norwalk, CT.

designed for multitasking, making it possible to monitor, acquire, and process data from both the TG and MS simultaneously. In the control mode the operator can usually set and adjust the parameters of both instruments from the keyboard. Most TG/MS systems are capable of continuous or disjointed spectrum scanning up to a predetermined atomic mass unit. While each TG/MS system differs somewhat, the general software and hardware features are summarized as follows:

- gas ratio analysis of the components of a mixture
- high sensitivity with regards to the detection of minor gas components
- keyboard control of major instrument parameters
- fast acquisition speeds in the range of 1000 amu/sec
- flexible, easily manipulated, and user interactive with graphic capability
- subtraction of background gases
- scaling, translation, and normalization of data
- capability of continuous or disjointed spectrum scanning up to a predetermined atomic mass unit

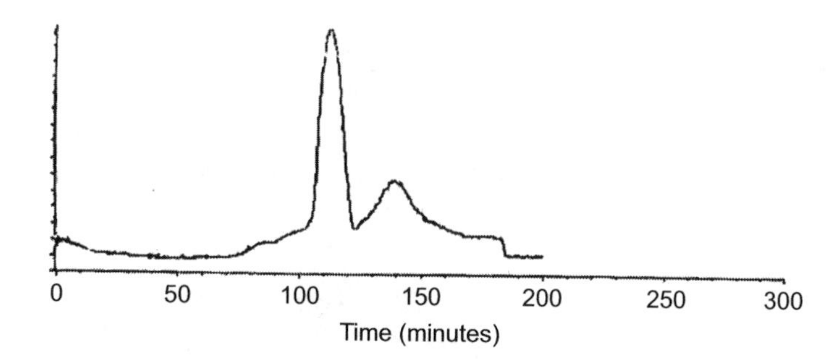

Figure 1.26 The m/z 28 mass chromatogram for the thermal decomposition of calcium oxalate monohydrate under high vacuum conditions (unpublished work of author).

- capability to overlay the TG curve on one or more of the mass chromatograms
- multiple ion monitoring in which several individual masses can be selected and displayed on the CRT
- library search of mass spectra data

Evolved gas data from the mass spectrometer include typical histogram spectra collected over the whole or part of the available mass range and/or mass chromatograms of specific masses (Fig. 1.26). The Total Ion Current (TIC) curve also may be obtained. Mass chromatograms reveal concentration changes of a particular mass over a designated time or temperature range, while TIC curves reveal concentration changes in total masses over the time or temperature range. TIC curves and TG curves often show close correlations. Overlaid TG and MS curves for the decomposition of $CuSO_4$ are shown in Fig. 1.27. As noted, mass spectral data may be correlated with gas evolutions data, and the temperature or time in which the gas evolutions take place may be determine as well as the identity of the gaseous components.

1.30.5 GC/MS, GC/FTIR and GC/MS/FTIR

The coupling of a gas chromatograph with a mass spectrometer and/or Fourier Transform Infrared Spectrophotometer is significant in that it allows the components of complex systems to be separated and identified. The sensitivity of these instruments is such that

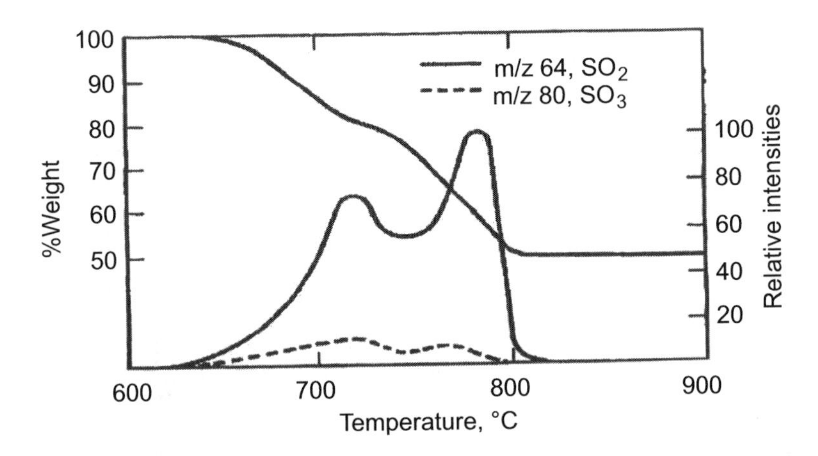

Figure 1.27 TG and MS profile curves for the decomposition of $CuSO_4$.

components with concentrations as small as a nanogram, can be separated, their spectra obtained, and their structure determined. The complementary structural information provided by simultaneous GC/MS/FTIR systems enhances component identification and is the principal advantage of GC/MS/FTIR over GC/MS or GC/FTIR.

As early as 1968, the analytical advantages of a combined GC/MS/FTIR system were discussed, but computer technology and software programming had not been sufficiently developed to make such systems practical. In the intervening years, many developments took place in these areas, but it was not until 1980 that a combined GC/MS/FTIR was developed. Presently, these systems may be purchased for less than $100,000, making them within financial reach of small colleges and universities. These systems are capable of rapidly analyzing complex mixtures from one injection into the GC.

Some manufacturers have developed combined GC/MS/FTIR systems configured in a parallel fashion. In this configuration, the GC effluent is divided, with a portion directed to the MS and the remaining portion directed to the FTIR. The serial configuration has the effluent passing through an infrared detector light pipe (Fig. 1.28) and from there to the mass spectrometer. One manufacturer has the transfer line to and from the IR light pipe and the transfer line to the mass spectrometer, all located in the oven

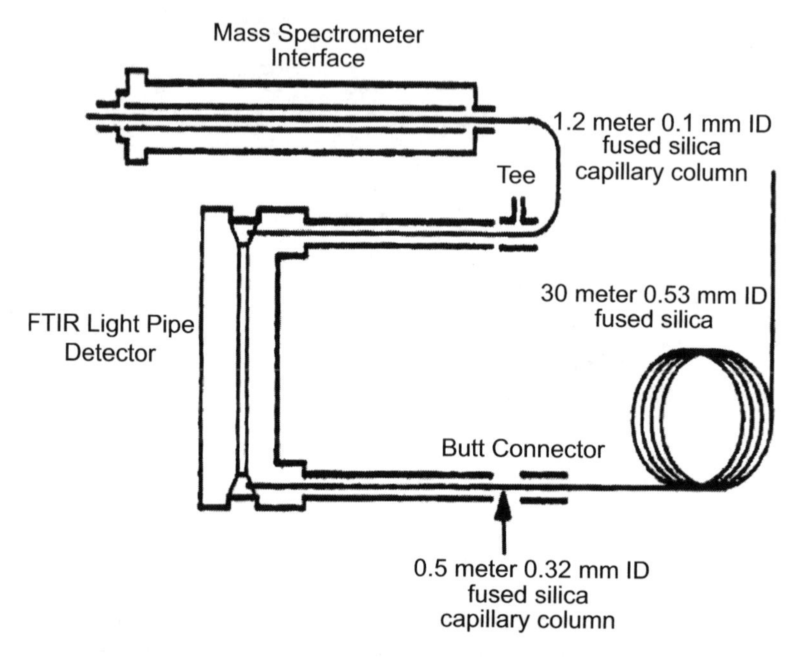

Figure 1.28 Schematic diagram of a serial configuration of a GC/MS/FTIR interface. *Source*: Leibrand R. J., *Am. Lab.*, December, 1988.

of the GC. According to the manufacturer, this arrangement allows the system to be configured in either a parallel or a serial manner. In addition, most of the systems may be operated as GC/FTIR or GC/MS.

1.30.6 The GC/MS and GC/FTIR and GC/MS/FTIR Interfaces

Again, the interface coupling the GC and MS must be designed to allow the components in the GC effluent to enter the chamber of the mass spectrometer without appreciably changing its vacuum. This is especially true for packed columns which are characterized by large amounts of carrier gases. This requirement can be accomplished with a jet spray, which is designed to partially remove the carrier gas (usually helium) before the GC effluent enters the chamber of the mass spectrometer. The jet spray (Fig. 1.29) which is surrounded

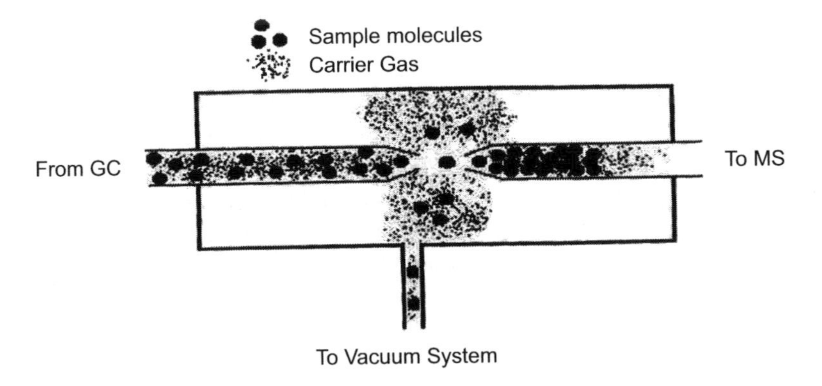

Figure 1.29 A schematic drawing of the jet spray GC/MS interface.

by a vacuum, is designed to remove a significant portion of the carrier gas as the GC effluent passes through a small orifice. After passing through the orifice, most of the carrier gas diffuses away (since it has a higher diffusion rate than the sample components). The components then pass into another chamber, which is coaxial to the first one. From there they enter the chamber of the mass spectrometer in a more concentrated form. Studies have shown that the jet spray device can remove about 90% of the carrier gas.

For combined capillary GC/MS/FTIR systems, the capillary column of the GC may be placed directed into the FTIR light pipe. The spectra of the gases in the light pipe are obtained and from there the GC effluent travels to the mass spectrometer through a smaller capillary tube.

The coupling of a gas chromatograph with a mass spectrometer is significant in that it allows the components of a complex system to be separated and identified. The sensitivity of these systems is such that components with concentrations as small as a nanogram, can be separated, their spectra obtained and their chemical structure determined. In the early years, the analytical advantages of a combined GC/MS system were discussed, but computer technology and software programming had not been sufficiently developed to make such systems practical. In the intervening years many developments took place in these areas, but it was not until around 1970 that a combined GC/MS instrument was developed. Presently,

GC/MS systems may be purchased at a cost within the reach of small colleges and universities. These systems are capable of rapidly analyzing complex mixture from one injection into the GC.

The importance of these combined systems is illustrated in the analysis of thyme, whose mass chromatogram is shown in Fig. 1.30. Thyme is a fragrant herb of the mint family. It has medicinal and culinary uses. The solution of thyme is passed through the capillary column of the gas chromatograph, and since the components of thyme have different affinities for the adsorption material that lines the walls of the capillary column, they exit the column at different times (called the retention times). The components then enter the mass spectrometer where they are bombarded by high-energy electrons, and the fragmentation pattern (mass spectrum) of each component is determined. With the proper software, the mass spectrum of each component can be searched against a library of mass spectra data and the structural identity of the components determined as shown in Fig. 1.31.

The analysis of thyme was conducted in the author's laboratory using the Wiley Registry of Mass Spectra Data. Currently, the Wiley Library has over 700,000 mass spectra. The NIST Library can also be searched. NIST is not as large as Wiley but is said to have mass spectra that is not in the Wiley Library. Scientific Instrument

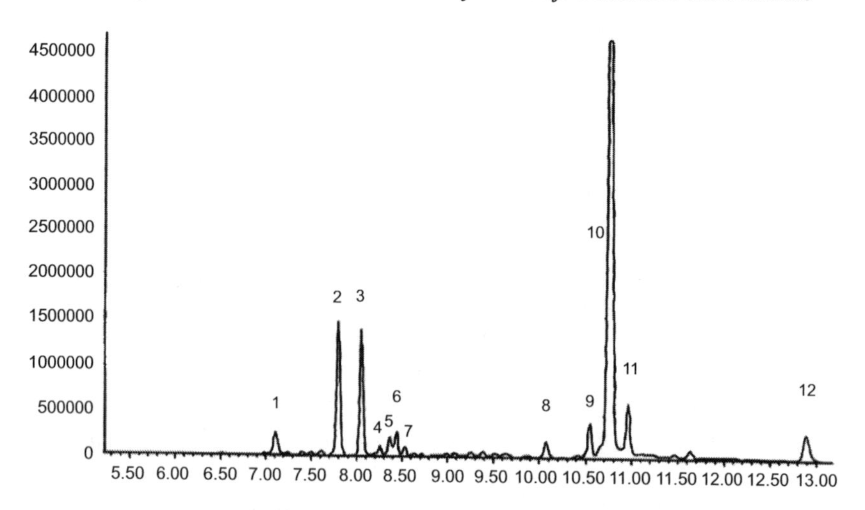

Figure 1.30 The mass chromatogram of thyme.

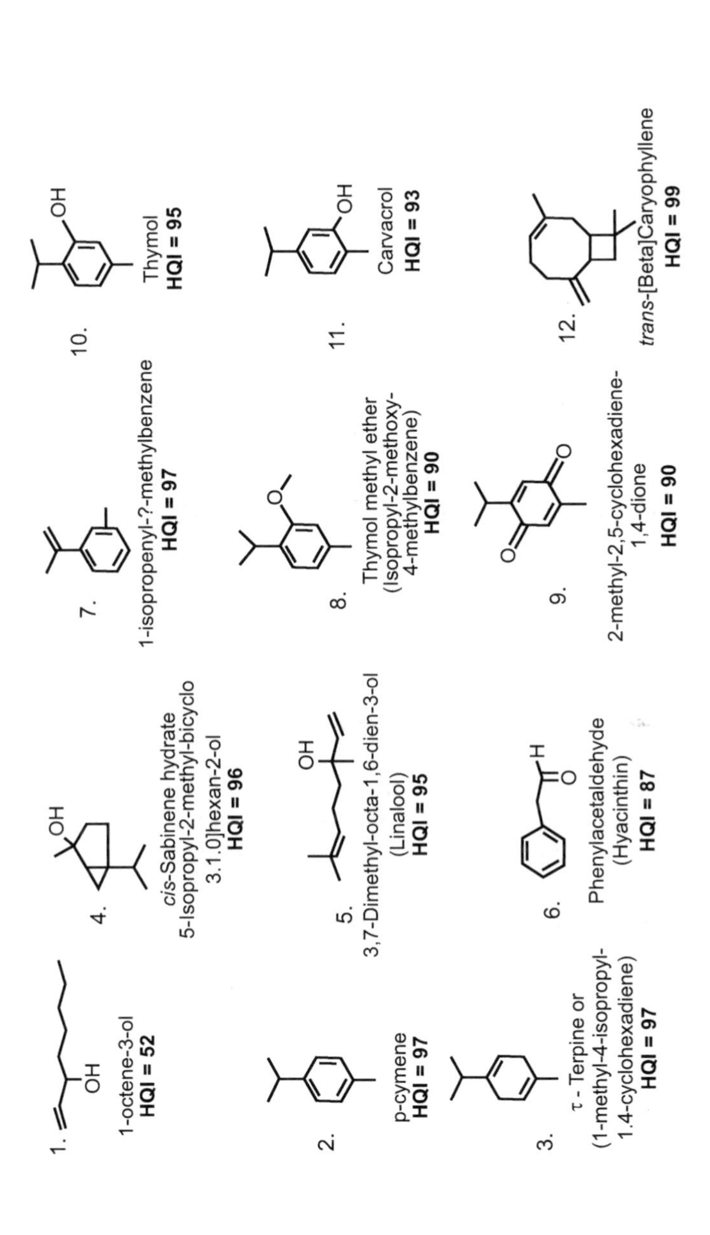

Figure 1.31 Organic compounds in thyme, *Note:* The Hit Quality Index (HQI) values were obtained by searching the mass spectrum of each of the components in the mass chromatogram against the Wiley Registry of Mass Spectra Data.

Services, the company that provides the Wiley Library, also has available the best of both worlds: a combined NIST/Wiley Library with all duplicate spectra removed. Also available from this company are a number of specialized mass spectra libraries, such as drugs, pesticides flavor, fragrances, designer drugs, and poisons.

References

1. Baumgartner E. and E. Nachbaur, *Thermochim. Acta*, **19**, 3 (1977).
2. Wilson D. E. and F. M., in *Thermal Analysis*, R. F. Schwenker and P. D. Garn, Editors, Academic, New York, vol. 1, p. 295 (1969).
3. Gibson E. K. and S. M. Johnson, *Thermochim. Acta.*, **4**, 49 (1972).
4. Chang T. L. and T. E. Mead, *Anal Chem.*, **43**, 534 (1971).
5. *Mettler Thermal Techniques*, Series T-107.
6. Langer H. G., R. S. Gohlke, and D. H. Smith, *Anal. Chim. Acta.*, **32**, 405 (1965).
7. Wiedemann, H. G., in *Thermal Analysis*, R. F. Schwenker and P. D. Garn, Editors, Academic, New York, vol. 1, p. 229 (1969).
8. Wendlandth W. W. and T. M. Southern, *Anal. Chim. Acta.*, **32**, 405 (1965).
9. Redfern, J. P., B. L. Treherne, M. L. Aspimal, and W. A. Wolstenholme, *17th Conference on Mass Spectrometry and Allied Topics*, Dallas, Texas, May 1969.
10. Gaulin C. A., F. Wachl, and T. H. Johnson, in *Thermal Analysis*, R. F. Schwenker and P. D. Garn, Editors, Academic, New York, vol. 2, p. 1453 (1969).
11. Dunner W. and H. Eppler, *Advanced Coupling Systems for Thermoanalyzers with Quadrupole Mass Spectrometers*, the 4th International Congress on Thermal Analysis (ICTA), Budapest, July 1974.

Chapter 2

The Analysis of Mass Spectra

2.1 The Fragmentation Patterns of Straight-Chain Alkanes

As a general rule, the spectra of straight-chain alkanes almost always show a molecular ion, but it is usually of weak intensity. Because of their smaller targets, low-molecular-mass alkanes will produce slightly more intense molecular ions than their higher-molecular-mass counterparts.

The preferred cleavage of pentane occurs in the following manner:

$CH_3CH_2CH_2CH_2$–|–CH_3 $CH_3CH_2CH_2$–|–CH_2CH_3

m/z 57   m/z 43

Major fragment   Minor fragment

The spectra of high-molecular-mass straight-chain alkanes (as well as other compounds containing long straight-chain hydrocarbon moieties) show a characteristic group of fragments appropriately referred to as the "hydrocarbon cluster." Each member of the cluster contains homologous fragments that differ from others by a mass of 14 (the mass of a CH_2 group) or multiples of 14 (Fig. 2.1).

Mass Spectrometry
James M. Thompson

ISBN 978-981-4774-77-2 (Hardcover), 978-1-351-20715-7 (eBook)
www.panstanford.com

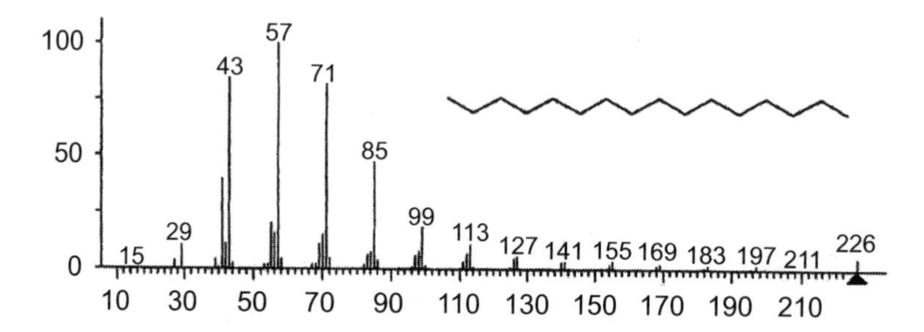

Figure 2.1 The spectrum of n-hexadecane, showing the "hydrocarbon cluster" beginning at m/z 29 $(CH_3CH_2)^+$.

High-molecular-weight alkanes also produce characteristic ions of relatively high abundances at m/z 71, 57, and 43. These ions correspond to $C_5H_{11}^+$, $C_4H_9^+$, and $C_3H_7^+$ respectively. A fragment ion at $(M - 15)^+$ is common to alkanes with branching methyl groups. However, the $(M - 15)^+$ ion is often of low intensity in straight-chain alkanes.

2.2 Branched Alkanes

The cleavage at branched carbons results in relatively stable fragments and is a preferred mode of fragmentation; therefore, as molecular branching increases, the intensity of M^+ decreases. The preferred fragmentation of branched alkanes is at the carbon leading to the most stable carbocation fragment. For instance,

$$CH_3-\overset{\displaystyle CH_3}{\underset{\displaystyle CH_3}{C}}{+}\ \cdot CH_2CH_3 \longrightarrow CH_3-\overset{\displaystyle CH_3}{\underset{\displaystyle CH_3}{C}}{+} \quad + \quad \cdot CH_2CH_3$$

2.3 Cycloalkanes

Non-branched cycloalkanes produce more intense M^+ ions than straight-chain alkanes of comparable molecular mass. This follows from the fact that the cleavage of one bond in straight-chain alkanes results in two fragments, while the cleavage of one bond in cycloalkanes keeps the mass of the molecule intact. Non-specific rearrangements are more prevalent among cycloalkanes than for straight-chain or branched chain alkanes, resulting in slightly more complex spectra.

As expected, the cleavage of substituted cycloalkanes is favored at the bond that connects the substituent to the ring (alpha-cleavage).

Cycloalkanes are also known to lose neutral ethylene, producing an $(M{-}C_2H_4)^+$ fragment, apparently by the following mechanism.

Other characteristic fragments are located at *m/z* 26, 27, 29, 41, 55, 69, and 83, some of which are discussed in the spectrum of methylcyclohexane shown in Fig. 2.2.

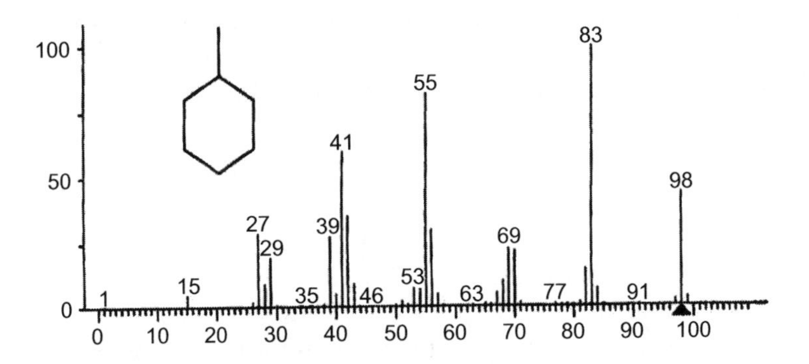

Figure 2.2 The spectrum of methylcyclohexane. *m/z* 98: M^+; *m/z* 83: $(M - CH_3)^+\alpha$-cleavage; *m/z* 70: $(M - C_2H_4)^+$ loss of ethylene; *m/z* 55: $(M - C_3H_7)^+$.

CH_3 → + CH_3

M^+ m/z 56

m/z 55: $\overset{+}{C}H_2CH_2CH{=}CH_2$

H — $-H\bullet$ →

M/z 56 M/z 55

M/z 29 $^+CH_2CH_3$

CH_3 → $H_2C{=}CH_2$ + CH_3 $-H\bullet$ → CH_3

M^+ + H H m/z 70 m/z 69

2.4 Unsaturated Hydrocarbons

The intensity of the molecular ion of unsaturated hydrocarbons decreases with branching and also with increasing molecular mass. Cleavage is always favored at an allylic carbon with the allylic carbon carrying the positive charge. This mode of fragmentation may be attributed to the relatively high stability of allylic ions.

$-\overset{|}{\underset{|}{C}}-\underset{|}{C}{=}C<$

Allylic carbon Vinylic carbons

The spectra of long-chain, terminal alkenes show the familiar "hydrocarbon cluster," which contains, among other fragments, a series of homologous fragments at *m/z* 41, 55, 69, etc., due to successive loss of CH_2 from $^+CH_2(CH_2)_nCH{=}CH_2$. Terminal alkenes often show intense fragments at m/z 27, which is probably generated by the following mechanism.

m/z 27

The base peak for cyclohexene appears at m/z 67, and its formation may be rationalized as involving a hydrogen abstraction followed by the loss of a methyl radical.

$+ \bullet CH_3$

m/z 67

Cyclohexene, as well as other cycloalkenes, undergoes what is commonly termed a retro Diels–Alders process, producing a relative intense fragment at m/z 54 as shown below. In most cases, the retro Diels–Alders fragmentation also produces a positive charged diene fragment and a neutral alkene. Some examples of the retro Diels–Alders fragmentation processes are shown below.

$\rightarrow H_2C{=}CH_2 +$

m/z 54

$+ H_2C{=}CH_2$

$+ H_2C{=}CH_2$

The molecular ions of aryl alkyls are stabilized by resonance and are, therefore, more intense than are the molecular ions of non-aromatic hydrocarbons. The cleavage of aryl alkyls is favored at the alpha carbon, producing the stabilized tropylium ion at m/z 91.

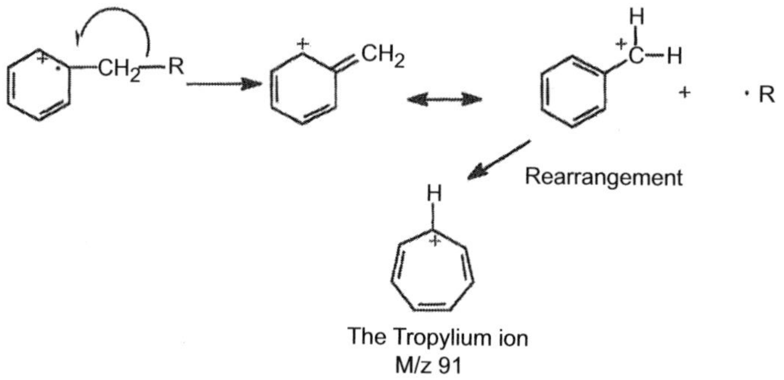

CH2—R → CH2 ↔ H +C—H + ·R

Rearrangement

H

The Tropylium ion
M/z 91

For mono-substituted aryl alkyls, the tropylium ion is often the base peak. If the aryl alkyl is disubstituted (such as in the isomeric xylenes), it will lose one of the alkyl groups to form a mono-substituted tropylium ion.

A favored cleavage of phenyl alkyls is at the bond between the alkyl group and the ring. This usually produces a fragment ion of strong to medium intensity at m/z 77, which along with other characteristic fragments at m/z 39, 50, 51, and 52 is taken as proof for the presence of the benzene ring. In addition to the fragments mentioned, benzene and mono substituted benzenes often produce low intensity fragments at m/z 78 and 79, which represent $(C_6H_6)^+$ and the ^{13}C isotope $(^{13}C^{12}C_5H_6)^+$ respectively (Fig. 2.3). To a lesser extent, the fragment at m/z 79 may be due to $(C_6H_6{+}H)^+$. The presence of these fragments affords additional proof that the compound contains a benzene ring.

In Fig. 2.3, the probable structures of the fragments at m/z 39 and 51 are shown below.

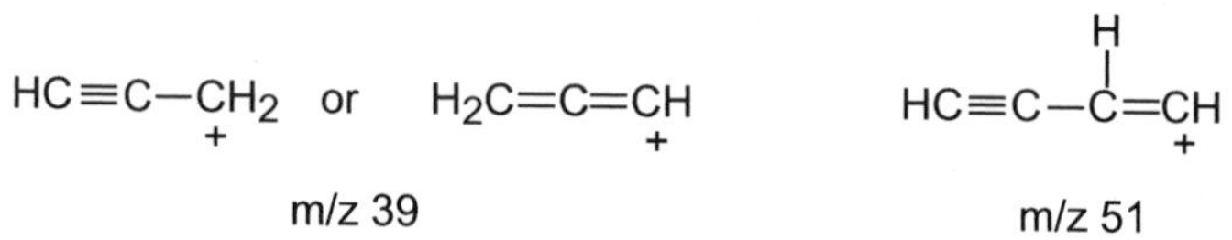

$HC{\equiv}C{-}\overset{}{\underset{+}{C}}H_2$ or $H_2C{=}C{=}\underset{+}{C}H$

m/z 39

$HC{\equiv}C{-}\overset{H}{C}{=}\underset{+}{C}H$

m/z 51

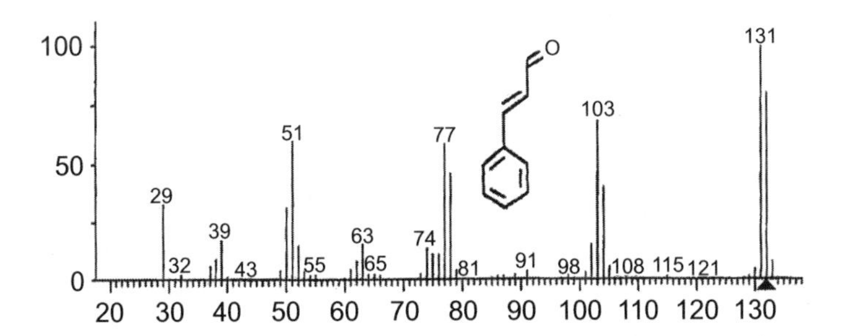

Figure 2.3 The mass spectrum of cinnamaldehyde, showing the series of fragments related to mono substituted benzenes. These fragments are located at m/z 39, 51, 77, 78, and 79.

2.5 Alkyl Halides

Low-molecular-weight chlorides, bromides, and iodides usually produce measurable but weak molecular ions (Figs. 2.4 and 2.5). Generally, the intensity of M^+ is more pronounced in the spectra of alkyl iodides and weakest or unobserved in the case of alkyl fluorides.

Alkyl bromides and chlorides also show characteristic isotopic clusters, whose appearance depend upon the number of bromines

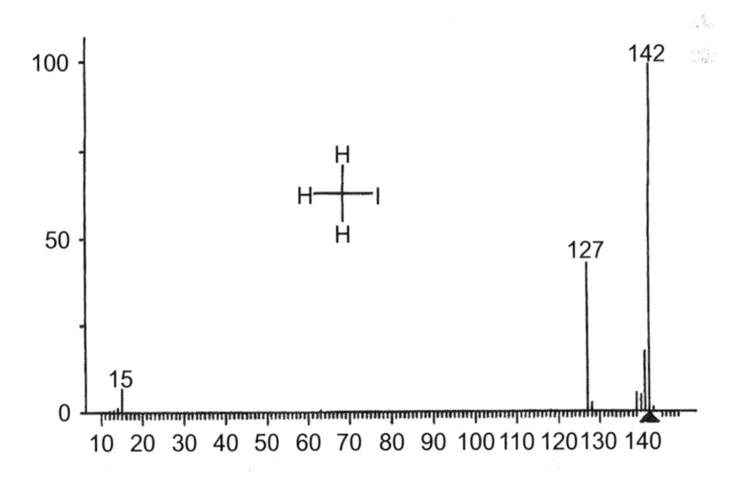

Figure 2.4 The mass spectrum of methyl iodine. m/z 142: M^+; m/z 141: $H_2C{=}I^+$; m/z 127: I^+; m/z 15: $CH_3{}^+$.

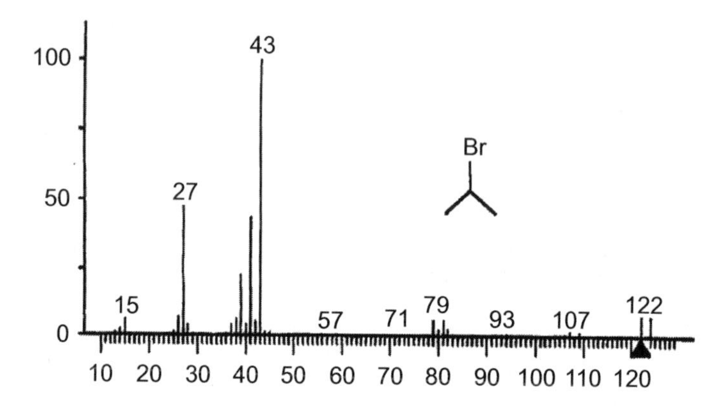

Figure 2.5 The mass spectrum of 2-bromopropane. *m/z* 122: M^+, $(C_3H_7{}^{79}Br)^+$; *m/z* 124: $(M + 2)^+$, the ^{81}Br isotopic peak $(C_3H_7{}^{81}Br)^+$ (this fragment is approximately 95% of the M^+ peak, indicating the presence of bromine); *m/z* 107: $(M - 15)^+$, probably due to $(CH_3)HC{=}Br^+$; *m/z* 82: $H^{81}Br^+$; *m/z* 80: $H^{79}Br^+$; *m/z* 81: $^{81}Br^+$; *m/z* 79: $^{79}Br^+$; *m/z* 43: Base peak, $(M{-}Br)^+$ or $(CH_3)_2CH^+$; *m/z* 41: Probably $CH_3{-}\overset{+}{C}{=}CH_2$; *m/z* 27: Probably $H_2C{=}\overset{+}{C}{-}H$.

and/or chlorines in the compound (Figs. 1.16, 1.17, and 1.18). The absence of an isotopic cluster in a suspected alkyl or aryl halide suggests the presence of monoisotopic fluorine or iodine, rather than bromine or chlorine. Also, if the $(M{+}1)^+/M^+$ ratio of a suspected alkyl halide is suspiciously small, an alkyl fluoride or iodide is suggested.

Cleavage of straight-chain alkyl halides occurs at the C–C bond adjacent to the halide, generating a weakly intense halonium ion as shown below. As noted, a similar cleavage is observed for straight-chain amines, alcohols, and thio-alcohols, forming the familiar immonium, oxonium, and thionium ions, respectively.

$$R{-}CH_2{-}\overset{\bullet +}{X} \longrightarrow H_2C{=}\overset{+}{X} + R\bullet$$

Where X = F, CL, Br, I, OH, SH and NH_2

The spectra of alkyl halides containing eight or more carbons tend to exhibit relative intense ion fragments, corresponding to $^+C_5H_{10}X$, $^+C_3H_6X$, and $^+C_4H_8X$, etc.

2.6 Phenyl Halides

Phenyl halides, without alkyl substituents, usually exhibit intense molecular ions. In addition to the characteristic isotopic cluster of phenyl bromides and chlorides, these compounds, as expected, produce the series of fragments common to the phenyl ring (i.e., m/z 79, 78, 77, 52, 51, and 39). In addition, the spectrum of phenyl halides often shows intense $(M\text{–}X)^+$ fragments.

2.7 Benzyl Halides

The molecular ion of benzyl halides, while generally not as intense as those of phenyl halides, is often of measurable intensity. A familiar characteristic of the spectrum of practically all benzyl compounds is the tropylium, located at m/z 91 (see Section 2.4).

2.8 Aliphatic Ethers

The molecular ions of aliphatic ethers are usually of weak intensity. Cleavage may occur at both the C–O and the C–C bond adjacent to the oxygen, producing fragments corresponding to RO^+ and $ROCH_2{}^+$ respectively. Some of the fragmentation mechanisms are shown below.

1. $RCH_2\text{—}\overset{+\bullet}{O}\text{—}R' \longrightarrow RCH_2^+ + {}^\bullet OR'$

2. $RCH_2\text{—}\overset{+\bullet}{O}\text{—}R' \longrightarrow {}^+R' + RCH_2O^\bullet$

3. $RCH_2\text{—}\overset{+\bullet}{O}\text{—}R' \longrightarrow R\dot{C}H_2 + \overset{+}{O}R$

4. $RCH_2\text{—}\overset{+\bullet}{O}\text{—}R' \longrightarrow {}^\bullet R + H_2C{=}\overset{+}{O}R$

5. etc.

The oxonium ion produced by reaction 4 above may undergo secondary fragmentation accompanied by the transfer of a hydrogen atom to give a characteristic fragment at m/z 31, as shown below with diethylether.

$$CH_3\text{–}CH_2\text{–}\overset{+\bullet}{O}\text{–}CH_2CH_3 \longrightarrow {}^{\bullet}CH_3 + H_2C{=}\overset{+}{O}\text{–}CH_2\overset{H}{C}H_2$$

$$\longrightarrow H_2C{=}\overset{+}{O}H + H_2C{=}CH_2$$

m/z 31

Cyclic ethers undergo fragmentation accompanied by the loss of formaldehyde.

$$\text{tetrahydropyran}^{+\bullet} \longrightarrow CH_2O + {}^{\bullet}CH_2CH_2CH_2CH_2^{+}$$

m/z 56

As a further example of the fragmentation patterns of aliphatic ethers, consider the spectrum of isopropyl-n-pentylether (Fig. 2.6).

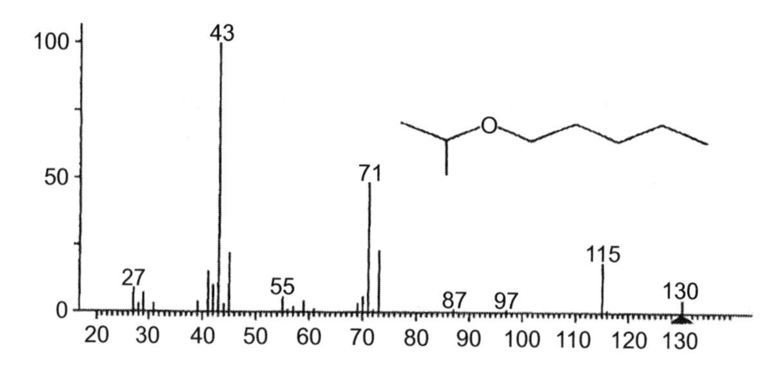

Figure 2.6 The mass spectrum of isopropyl-n-pentylether. *m/z* 130: M^{+}; *m/z* 115: $(M\text{–}CH_3)^{+}$; *m/z* 73: $H_2C{=}\overset{+}{O}CH(CH_3)_2$; *m/z* 71: $\overset{+}{C}H_2CH_2CH_2CH_2CH_3$; *m/z* 45: $CH_3CH{=}\overset{+}{O}H$; *m/z* 43: $\overset{+}{C}H(CH_3)_2$, $\overset{+}{C}H_2CH_2CH_3$ and $C_2H_3\overset{+}{O}$ (base peak); *m/z* 31: $H_2C{=}\overset{+}{O}H$; *m/z* 29: $\overset{+}{C}H_2CH_3$.

The mass spectrum of anisole shows weak but characteristic aromatic fragments at *m/z* 77 $(M\text{–}OCH_3)^{+}$, *m/z* 93 $(M\text{–}CH_3)^{+}$, *m/z* 65 $(M\text{–}CH_3)^{+}\text{–}CO$, *m/z* 78 $(M\text{–}CH_2O)^{+}$, and *m/z* 79 $(M\text{–}CHO)^{+}$. Aryl alkyl ethers containing an alkyl moiety of two or more carbons tend to lose $CH_2{=}CHR$, to give a fragment at *m/z* 94. A probable mechanism for the formation of this fragment may be visualized as follows:

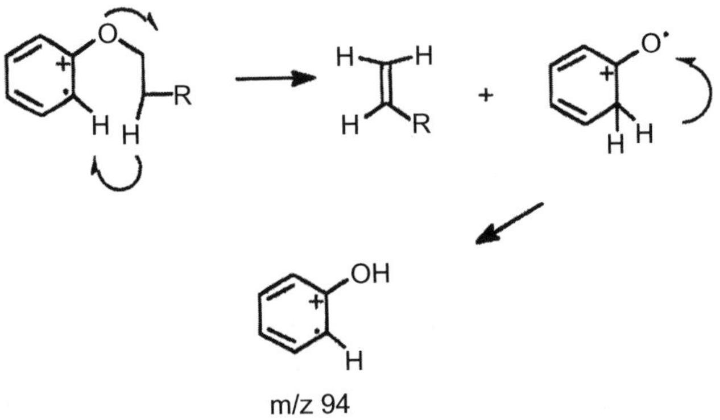

m/z 94

2.9 Alcohols

The mass spectra of primary alcohols are characterized by (M–H_2O) fragments formed by the loss of water from the molecular ion: For secondary and tertiary alcohols, the tendency to lose water is not quite as common. Because of the ease of dehydration, alcohols containing more than five carbons seldom exhibit molecular ions; however, this deficiency is often compensated by the tendency of the alcohol to form $(M - 1)^+$ ions through the loss of an alpha-hydrogen.

Primary, secondary, and tertiary alcohols all undergo cleavage to form the resonance stabilized oxonium. If the alcohol is primary, the ion appears as a conspicuous peak at m/z 31.

Low-molecular-weight amines and ethers also undergo alpha-cleavage to produce immonium and oxonium ions, which are among the more intense fragment in the spectra of these compounds.

As the molecular weight of the alcohol increases, so does competitive fragmentation. This results in a decrease in the intensity of the oxonium ion. In cases where the alpha-carbon is substituted, the oxonium ion may undergo further fragmentation to produce additional ions. An understanding of the fragmentation patterns of alcohols may be obtained from a consideration of the spectrum of 3-methyl-3-heptanol (Fig. 2.7).

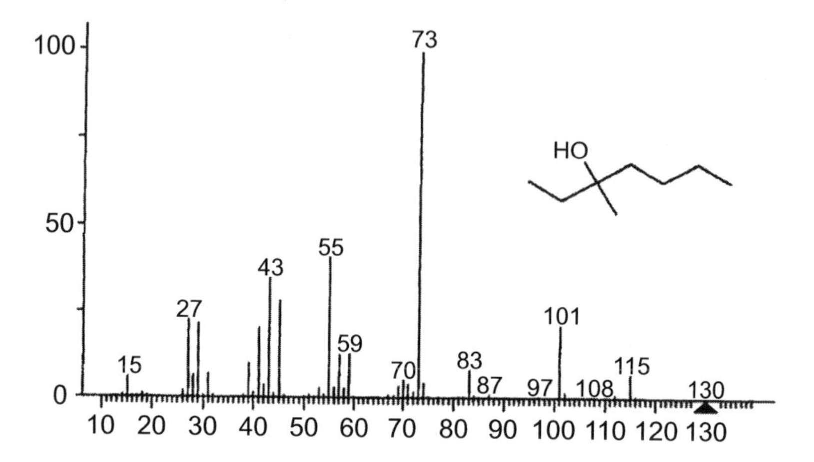

Figure 2.7 The mass spectrum of 3-methyl-3-heptanol.

Also,

$$CH_3CH_2CH_2CH_2-\underset{C_2H_5}{\overset{CH_3}{\underset{|}{\overset{|}{C}}}}-\overset{\bullet +}{O}H \longrightarrow \bullet C_2H_5 + CH_3CH_2\overset{H}{\overset{|}{C}}H-CH_2-\overset{CH_3}{\overset{|}{C}}=\overset{+}{O}H$$

m/z 101

$$\swarrow$$

$$CH_3CH_2CH=CH_2 \quad + \quad H\overset{CH_3}{\overset{|}{C}}=\underset{+}{O}H$$

m/z 45

Also,

$$CH_3CH_2CH_2CH_2-\underset{C_2H_5}{\overset{CH_3}{\underset{|}{\overset{|}{C}}}}-\overset{\bullet +}{O}H \longrightarrow \bullet CH_3 + CH_3CH_2\overset{H}{\overset{|}{C}}H-CH_2-\underset{C_2H_5}{\underset{|}{C}}=\overset{+}{O}H$$

m/z 115

m/z 83: Probably occurs from the loss of water from m/z 101.

m/z: 73: $CH_3-\underset{C_2H_5}{C}=\overset{+}{O}H$ (base peak)

m/z: 59: $CH_3CH_2CH=\overset{+}{O}H$

$$CH_3CH_2\underset{H}{\overset{H}{C}}-CH_2\underset{CH_2CH_3}{C}=\overset{+}{O}H \longrightarrow CH_3CH_2CH=CH_2 + \underset{CH_2CH_3}{CH}=\overset{+}{O}H$$

m/z 115 m/z 59

m/z 55: probably due to m/z $(73\text{-}H_2O)^+$
m/z 45: $CH_3CH=\overset{+}{O}H$, see the mechanism above.
m/z 48: probably $CH_3C\equiv\overset{+}{O}$ or $H_2C=C=\overset{+}{O}H$ and $\overset{+}{C}_3H_7$
m/z 29: $HC\equiv\overset{+}{O}$ and/or $\overset{+}{C}_2H_5$

The mass spectrum of 2-methylcyclohexanol shows a base peak at *m*/*z* 57, which is believed to be formed by the following mechanism:

$\overset{\bullet +}{O}H$ CH$_3$ → $\overset{+}{O}H$ CH$_3$ → $\overset{+}{O}H$ CH$_3$

$\overset{+}{O}H$ + CH$_3$

m/z 57

Other major fragments are located at m/z 114 (M^+), 86, 71, and 57. The probable mechanism for the formation of these ions is shown below:

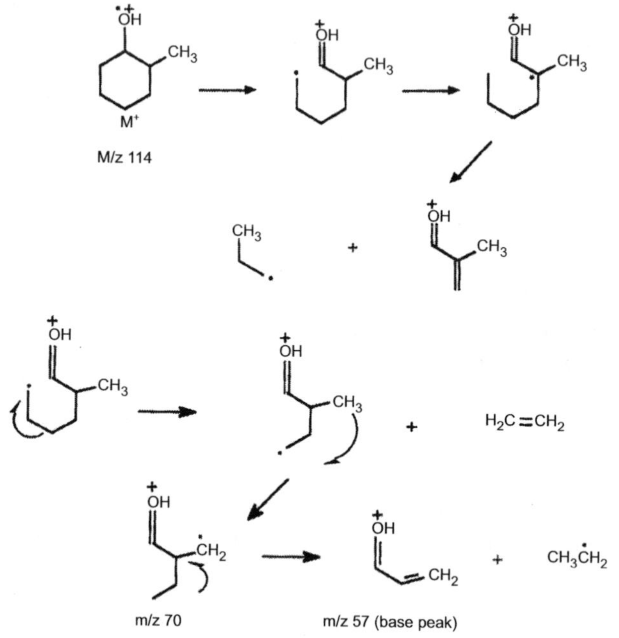
OH
CH3
M+
M/z 114
H2C=CH2
CH3CH2
CH2
m/z 70
m/z 57 (base peak)

Other significant fragments found in the spectrum of 2-methylcyclohexanol are located at m/z 96 (M – H_2O) and m/z 44. A plausible mechanism for the formation of the latter fragment is as follows:

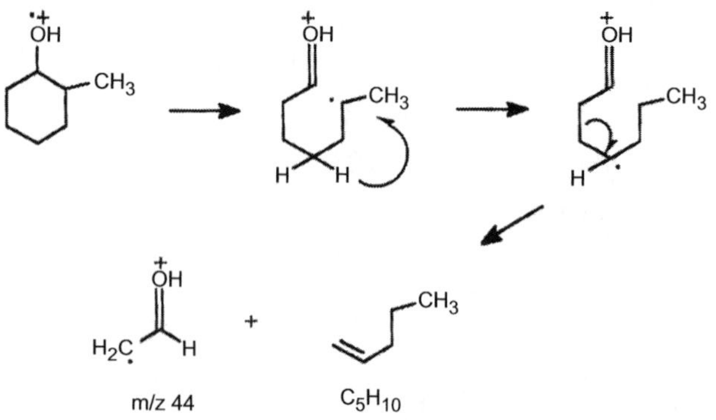
OH
CH3
H
H2C
m/z 44
C5H10

2.10 Phenols

The spectra of phenols exhibit intense molecular ions and characteristic fragments at $(M - CO)^+$ and $(M - CHO)^+$.

2.11 Aliphatic Ketones

Aliphatic ketones usually produce measurable molecular ions. The cleavage is preferred at an alpha-bond giving rise to the resonance stabilized acylium ion which often constitutes the base peak. If the ketone has a methyl carbonyl group, then the base peak is often the methyl acylium ion located at m/z 43.

$$R{-}C(R'){=}O^{\bullet+} \longrightarrow {}^{\bullet}R' + R{-}C{\equiv}O^+$$

The acylium ion
if R= CH_3, m/z = 43

In the formation of the acylium ion, aliphatic ketones prefer cleavage of the largest alkyl group. However, both R and R′ acylium ions are observed. The secondary fragmentation of $R\text{–}C{\equiv}O^+$ results in the formation of R^+ and neutral CO. An anomalous fragment located at $(M\text{–}H_2O)^+$ is sometimes observed in the mass spectra of aliphatic ketones. Apparently, this fragment is formed by a nonspecific rearrangement.

Aliphatic ketones with alkyl groups of three or more carbons will undergo a typical McLafferty rearrangement, which involves the migration of a hydrogen atom to a carbonyl oxygen. This process is followed by the elimination of an olefin to form a fragment at m/z $(43 + R)^+$.

R′–CH(H) … C(=O•+)R → R′–ĊH … C(=O+H)R → R′–CH=CH2 + •CH2–C(=O+H)R

m/z (43+R)

2.12 Aromatic Ketones

The molecular ions of aromatic ketones are stabilized by resonance and often of considerable intensity. The base peak is usually due to the resonance stabilized aromatic acylium ion as shown below. In most cases, this ion undergoes additional fragmentation producing the phenyl ion at m/z 77 (assuming only mono-substitution of the aromatic ring) and CO.

For arylalkyl ketones, usually the aromatic and alkyl acylium ions are produced (as shown below), but the latter is of much weaker intensity. In the case of diaryl ketones, the same is true in that both acylium ions are formed.

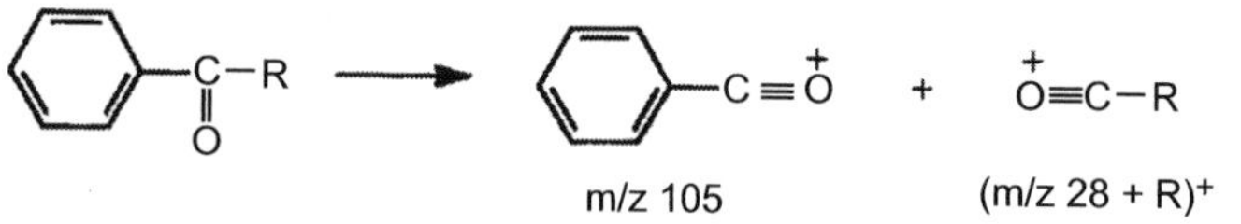

m/z 105 (m/z 28 + R)+

If the aromatic ketone contains a side chain of four carbons, the McLafferty rearrangement is observed as shown below (also see Section 1.21).

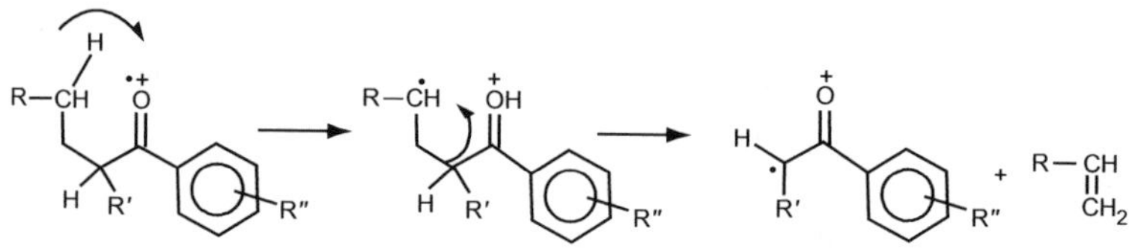

2.13 Aliphatic Aldehydes

The molecular ion of these compounds is usually observable but decreases with branching and with increasing molecular weight. Above four carbons atoms per molecule, the intensity of M^+ undergoes a rapid decrease in intensity.

Alpha cleavage to produce CHO^+ at m/z 29 is a dominate mode of fragmentation; however, if the aldehyde contains more than four carbons, the intensity of the m/z 29 fragment is enhanced by overlap with the $^+C_2H_5$ ion.

The oxonium ion is a common fragment in the spectra of these compounds and is often the base peak. For aldehydes that are unbranched at the alpha carbon, the oxonium ion appears at m/z 44 and is probably the result of a McLafferty rearrangement similar to the mechanism that follows:

m/z 44 (R′ = H)

The mass spectra of aliphatic aldehydes also show a significant $(M\text{–}H)^+$ fragment, which results from the loss of an aldehydic hydrogen. Other characteristic fragment ions of aliphatic aldehydes are discussed in the spectrum of propionaldehyde (Fig. 2.8).

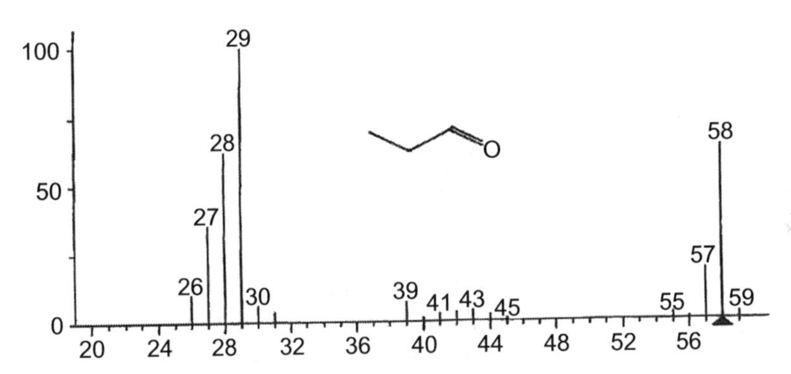

Figure 2.8 The spectrum of propionaldehyde. m/z 59: $(M + 1)^+$, due to the ^{13}C isotope; m/z: 58 M^+; m/z 57: $(M - H)^+$, due to the loss of an aldehydic hydrogen; m/z 43: $(CH_2CHO)^+$; m/z 31: $(H_2C{=}OH)^+$; m/z 30: $H_2C{=}O^+$; m/z 29: $CH_3CH_2^+$ and $H\text{–}C{\equiv}\overset{+}{O}$.

2.14 Aromatic Aldehydes

If the carbonyl group is attached directly to the ring, as in the case of benzaldehyde, the molecular ion is usually intense.

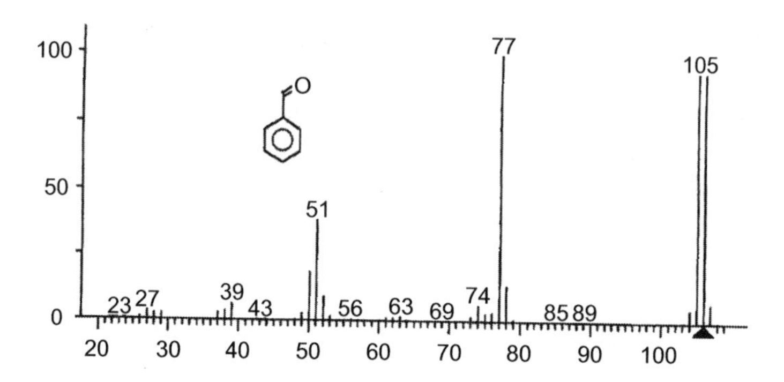

Figure 2.9 The mass spectrum of benzaldehyde. *m/z* 106: M^+; *m/z* 105: $(M\text{–}H)^+$, loss of an aldehydic hydrogen; *m/z* 78: (M–28)+, loss of CO; *m/z* 77: $(M\text{–}CHO)^+$, base peak, due to the $^+C_6H_5$ group; *m/z* 76: $(M - CH_2O)^+$; *m/z* 51: $[(M - CO) - HC{=}CH_2)]^+$ or $[(M - CHO) - HC{\equiv}CH)]^+$.

Again, a preferred mode of fragmentation for these compounds involves the loss of the aldehydic hydrogen producing intense $(M\text{–}H)^+$ ions. Other major fragmentation patterns may be illustrated in the spectrum of benzaldehyde shown in Fig. 2.9.

In Fig. 2.9, it should be noted that the intensity of the $(M\text{–}H)^+$ fragment at *m/z* 105 is equal in intensity to M^+. This attests to the stability of the aromatic acylium ion $(ArC{=}O)^+$, which is a characteristic fragment of these compounds. Also typical of the spectra of aromatic aldehydes is a fragment corresponding to the loss of CO. For benzaldehyde, shown in Fig. 2.9, the fragment is located at *m/z* 78. The base peak located at *m/z* 77, in Fig. 2.9, is due to the loss of CHO from the molecular ion, and its intensity will vary depending upon the number and nature of the substituents on the ring. The fragment at *m/z* 77 appears to undergo further fragmentation, losing HC=CH to produce the ion at *m/z* 51.

To calculate the carbon number for benzaldehyde, assuming it is not known, we consider the abundance of fragments M^+ and $(M + 1)^+$ which is 994 and 72, respectively. These fragments are adjusted to 100% abundance by dividing 994 (M^+) into itself and into 72 $(M + 1)^+$, when the results are multiplied by 100 values of 100 and 7.627 are obtained. Dividing 7.627 by 1.08 (the natural abundance of the ^{13}C isotope), 7.06 is calculated for the carbon number.

2.15 Anhydrides

Molecular ions are usually absent or of weak intensity in the spectra of open chain aliphatic anhydrides; however, intense $(M + H)^+$ ions are often produced at higher pressures. Alkyl acylium ions $(R\text{–}C{\equiv}O^+)$ are fragments of anhydride. For instance, the base peak for acetic anhydride is $CH_3\text{–}C{\equiv}O^+$, which is located at m/z 43. Another common fragmentation mode for anyhdride is the loss of CO from the acylium ion to form R^+. As an example, consider the fragmentation of butyric anhydride:

$$CH_3CH_2CH_2\text{–}C(=O)\text{–}O\text{–}C(=O)\text{–}CH_2CH_2CH_3 \longrightarrow CH_3CH_2CH_2\text{–}C{\equiv}\overset{+}{O}$$

m/z 71 (base peak)

-CO

$CH_3CH_2CH_2$

m/z 43 (second most intense peak)

Cyclic aliphatic anhydrides often produce fragments corresponding to $(M\text{–}CO)^+$; however, the latter fragment is less common.

2.16 Acetals and Ketals

These compounds usually do not produce measurable molecular ions, but in the case of acetals, conspicuous $(M + 1)^+$ are produced as a result of the loss of the tertiary hydrogen. The formation of R^+ and OR^+ by cleavage of the alpha-bond is also a favorite fragmentation pattern. The cleavage at the C–O bond is preferred over the C–C cleavage; consequently, the base peak generally corresponds to the loss of the ^+OR fragment.

$$H\text{–}C(OR')(R)\text{–}\overset{\bullet +}{O}R'' \longrightarrow {\cdot}R + (R'O)(H)C{=}\overset{+}{O}R''$$

$$H\text{–}C(OR')(R)\text{–}\overset{\bullet +}{O}R'' \longrightarrow {\cdot}OR' + (R)(H)C{=}\overset{+}{O}R''$$

2.17 Aliphatic Acids

Straight-chain monocarboxylic acids produce weak molecular ions whose intensities decrease even further with increasing molecular weight. Aliphatic acids containing four or more carbons undergo the McLafferty rearrangement, resulting in the formation of an ion fragment at m/z 60. Depending upon the structure of the acid, the intensity of the m/z 60 fragment may vary from negligible to the base peak. If the acid is branched at the α-carbon, the McLafferty rearrangement produces a fragment at m/z $(59 + R)^+$.

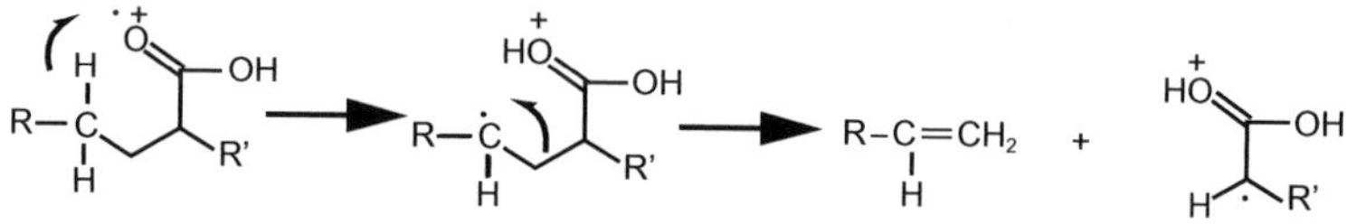

R–C=CH$_2$ +

m/z 60, If R′ = H

Other fragments common to organic acid include the following:

(1) $(COOH)^+$, m/z 45
(2) $(M\text{–}OH)^+$
(3) $(M\text{–}COOH)^+$
(4) $(M\text{–}H_2O)^+$

Ions 2, 3, and 4 are typical of low-molecular-weight acids and may not be observed in the mass spectra of higher molecular weighs acids.

Large alkyl moieties of straight-chain acids undergo fragmentations reminiscent of long straight-chain alkanes, producing the familiar "hydrocarbon cluster" (see Fig. 2.1). Each cluster member differs from those of proceeding members by mass units of 14, which is attributed to successive CH_2 groups. Additional cluster peaks common to the spectra of high-molecular-mass straight-chain acids include $^+CH_2COOH$, $^+CH_2CH_2COOH$, $^+CH_2CH_2CH_2COOH$, etc.

Dicarboxylic acids, because of their low volatility, usually do not produce representable spectra, but the problem may be resolved by preparing diester derivatives.

2.18 Aromatic Acids

Monocarboxylic aromatic acids produce relatively intense molecular ions as a result of resonance stabilization. As in the case of aliphatic acids, these compounds show fragment ions corresponding to $(M–OH)^+$, $(M–H_2O)^+$, and $(M–COOH)^+$. The fragment at $(M–H_2O)^+$ is especially prominent in the spectra of acids containing an *ortho* substituent that has a labile hydrogen (see below). From all indications, the loss of water among acids is facilitated by the ability of the acid to form six-membered transition states made possible by the labile hydrogen.

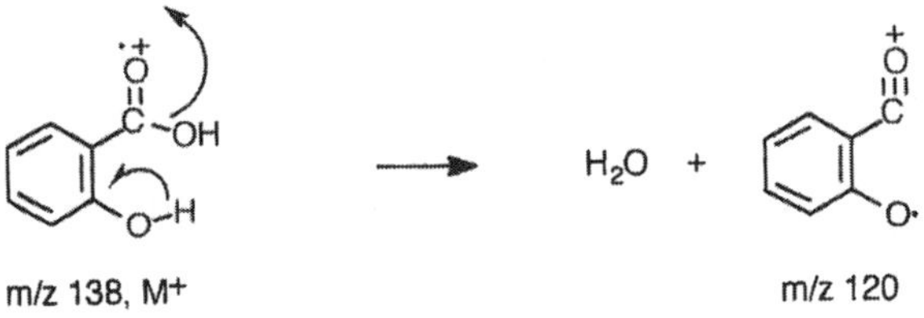

m/z 138, M^+ → H_2O + m/z 120

The above mechanism is an example of the so-called "*ortho* effect" and is observed in a variety of structures conforming to the following general structure:

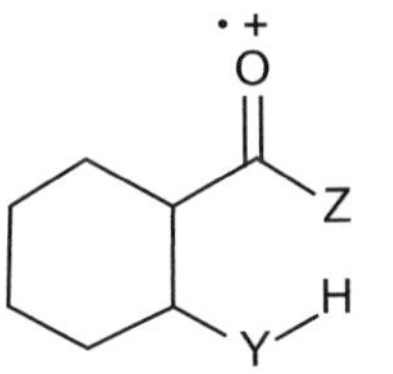

Where: Z = OH, OR or NH_2
Y = CH_2, O or NH

2.19 Aliphatic Amines

If the structure of an amine reveals only one nitrogen, the molecular ion of the amine must have an odd numerical value. However, in the case of long-chain amines, M^+ is usually not measurable and for low-molecular-weight amines, it is often of weak intensity. The spectra of amines are similar, in some respects, to ethers and methyl esters in that the cleavage is favored at the bond between the carbon and the heteroatom. In aliphatic amines, it is usually this cleavage that produces the base peak. If the amine is branched at the alpha-carbon, one would also expect the cleavage to occur at the point of branching. Under relatively moderate sample pressure, there is a tendency for amines to form $(M + H)^+$ ions, a consequence of the ease in which nitrogen is protonated through molecule-ion collisions. The importance of identifying $(M + H)^+$ cannot be overstressed, especially in cases where no molecular ion is present.

Straight-chain primary amines show relatively abundant fragments at m/z 30, which correspond to the following immonium ion:

$$\mathrm{H_2C = \overset{+}{N}H_2}$$
$$\mathrm{m/z\ 30}$$

For secondary amines, the fragment at m/z 30 is replaced by one at m/z (29 + mass of the substituent). For instance, the two most abundant fragments in the mass spectrum of 3-aminohexane are located at m/z 58 and 72. These masses correspond to the following fragments:

$$\mathrm{CH_3CH_2CH = \overset{+}{N}H_2} \qquad \mathrm{CH_3CH_2CH_2CH = \overset{+}{N}H_2}$$
$$\mathrm{m/z\ 58\ (base\ peak)} \qquad \mathrm{m/z\ 72}$$

The mechanism for the formation of an alkyl immonium ion is illustrated as follows:

$$\mathrm{R-\overset{\overset{+\bullet}{N}H_2}{\underset{H}{C}}-R'} \longrightarrow \mathrm{R(H)C=\overset{+}{N}H_2} + \mathrm{\bullet R'}$$
$$\mathrm{m/z\ 29 + R}$$

$$\mathrm{R-\overset{\overset{+\bullet}{N}H_2}{\underset{H}{C}}-R'} \longrightarrow \mathrm{R'(H)C=\overset{+}{N}H_2} + \mathrm{\bullet R}$$
$$\mathrm{m/z\ 29 + R'}$$

An example of the fragmentation process for tertiary amines is illustrated as follows:

$$CH_3CH_2CH_2-\overset{\overset{\Large CH_2CH_3}{|}}{\underset{\underset{\Large \overset{+}{N}H_2}{|}}{C}}-CH_3 \longrightarrow \left\{\begin{array}{ll} CH_3CH_2CH_2-\overset{\overset{\Large CH_2CH_3}{|}}{C}{=}\overset{+}{N}H_3 \;\; (m/z\ 100) & + \ \cdot CH_3 \\ CH_3CH_2CH_2-\overset{\overset{\Large CH_3}{|}}{C}{=}\overset{+}{N}H_2 \;\; (m/z\ 86) & + \ \cdot CH_3CH_2 \\ CH_3CH_2-\overset{\overset{\Large CH_3}{|}}{C}{=}\overset{+}{N}H_2 \;\; (m/z\ 72) & + \ CH_3CH_2\dot{C}H_2 \end{array}\right.$$

3-methyl-3-aminohexane

In this example, the ion at m/z 72 is the most intense of the three fragments, again reflecting the preferred loss of the highest mass substituent.

Some fragment ions common to the mass spectrum of diethylamines are shown in Fig. 2.10.

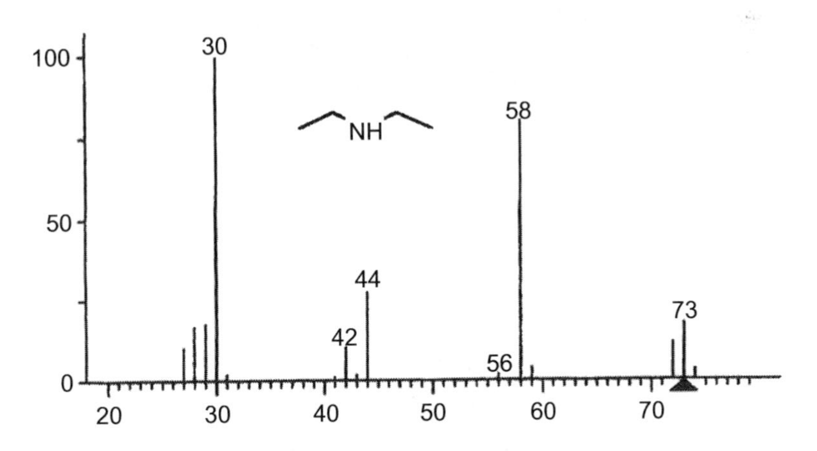

Figure 2.10 The mass spectrum of diethylamine.

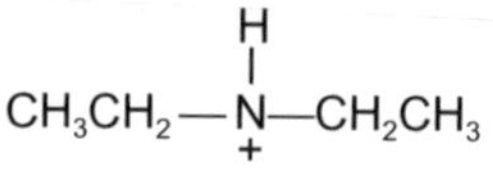

$CH_3CH_2-\overset{H}{\underset{+}{N}}-CH_2CH_3$

M^+

m/z 73

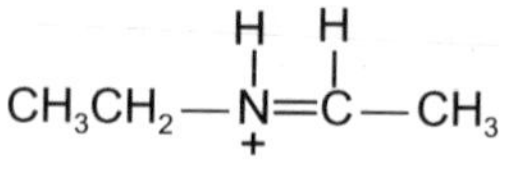

$CH_3CH_2-\overset{H}{\underset{+}{N}}=\overset{H}{C}-CH_3$

$(M-1)^+$

m/z 72

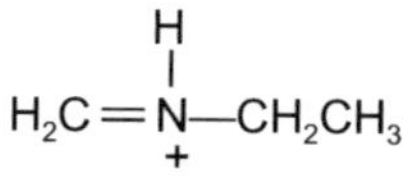

$H_2C=\overset{H}{\underset{+}{N}}-CH_2CH_3$

m/z 58

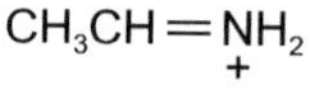

$CH_3CH=\underset{+}{N}H_2$

m/z 44

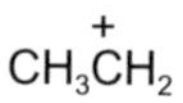

$CH_3\overset{+}{C}H_2$

m/z 29

$H_3C=\underset{+}{N}H_2$

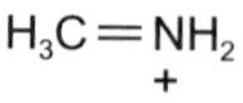

m/z 30 (base peak)

2.20 Aromatic Amines

Aromatic amines containing one nitrogen atom exhibit intense molecular ions with odd mass values. The cleavage at the C–C bond adjacent to nitrogen represents a major mode of fragmentation, giving rise to a fragment at m/z 106. If the aromatic ring contains another substituent, the fragment ion will be less intense and will appear at m/z $(105 + R)^+$, where R is the substituent.

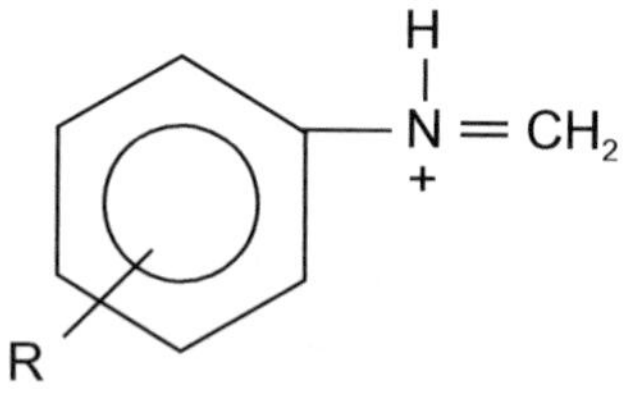

H

N = CH_2

\+

R

m/z 106 (R = H)
m/z 120 (R = CH_3), etc.

Aromatic amines also exhibit ion fragments corresponding to $(M\text{–}HCN)^+$.

2.21 Amides

These compounds usually produce weak molecular ions. For straight-chain primary amides containing four or more carbons in the chain, the McLafferty rearrangement results in a characteristic fragment ion at m/z 59; however, if the amide is branched at the alpha-carbon, the mass of the fragment will be increased to reflect the mass of the branched substituent. The fragment at m/z 59 appears in the spectrum of 3-methylbutyramide as the base peak, and its formation probably occurs by the McLafferty rearrangement as shown below:

3-methylbutyramide

m/z 59 (base peak)

The cleavage of amides also occurs at the alpha-bond, producing a characteristic fragment at m/z 44. The ion fragments at m/z 44 and 59 are shown in the spectrum of hexamide (Fig. 2.11).

$RCH_2-C(=O^{\cdot +})-NH_2 \longrightarrow R\dot{C}H_2 + {}^{+}O\equiv C-NH_2$

m/z 44

Primary straight-chain amides, containing an alkyl group of five or more carbons, may exhibit an ion fragment at m/z 86 due to cleavage of the gamma-bond (Fig. 2.11).

m/z 86

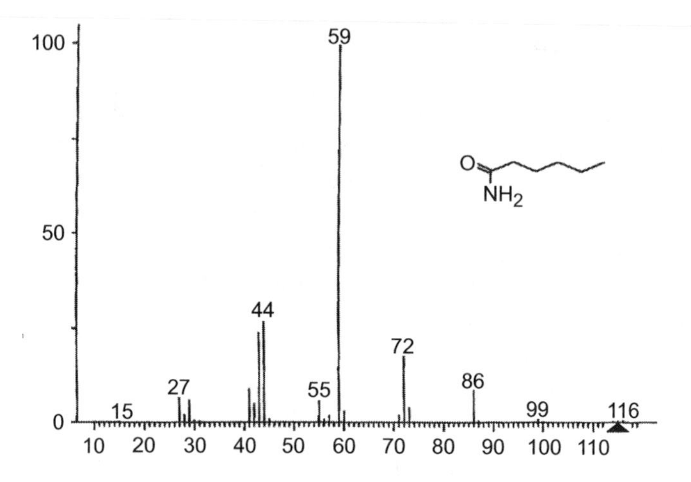

Figure 2.11 The mass spectrum of hexamide.

2.22 Aliphatic Nitro Compounds

Molecular ions of aliphatic nitro compounds are usually of weak intensity unless it is of fairly low molecular weight. Three fragments common to these compounds are as follows:

$$(M - NO_2)^+$$
$$(NO)^+, m/z\ 30$$
$$(NO_2)^+, m/z\ 46$$

2.23 Aromatic Nitro Compounds

Molecular ions of aromatic nitro compounds are usually stabilized by resonance and, therefore, of considerable intensity. Some of the characteristic fragments of these compounds are shown below using nitrobenzene as the example.

(1) [M–NO], base peak for nitrobenzene

(2) $\left[\text{cyclohexa-2,5-dienone} \right]^+$, a fragment common to mononitro aromatic compounds

(3) $[M–NO_2]^+$, for nitrobenzene, this fragment appears at m/z 77

(4) $[M–NO_2]–[HC{\equiv}CH]$, for nitrobenzene

Chapter 3

Problems in Mass Spectrometry

3.1 Introduction

The mass spectra that follow, for the most part, represent common organic compounds. The spectra have been selected to give the student experience in analyzing a broad class of organic compounds with a variety of structural features. In most cases, the mass spectra are presented with some supportive structural information such as the chemical formula. The answers to certain selected problems are listed at the end of this section.

In identifying the unknown structures, it is suggested that the student try to identify as many fragment ions as possible. Without other information (NMR, IR, UV spectra, and chemical formula), this suggested procedure may be insufficient for identifying the unknown compound. However, this procedure will increase the students' knowledge of the fragmentation procedure. Once the student has arrived at a possible structure, he or she may wish to compare the unknown mass spectrum with known spectra, thus providing absolute structure verification. Regarding this latter suggestion, it should be emphasized that the relative abundances of mass fragments for the same compound often vary among different mass spectrometers. Nevertheless, if the compounds are identical

Mass Spectrometry
James M. Thompson

ISBN 978-981-4774-77-2 (Hardcover), 978-1-351-20715-7 (eBook)
www.panstanford.com

and the mass spectra are obtained under the same conditions, there should be sufficient similarity to suggest identical structures.

The interpretation of organic spectral data follows no fixed procedure. The approach will depend upon the kinds of information desired and the unique style of the analyst. In interpreting organic spectra, every analyst will draw upon the fundamentals that have been discussed. Many suggestions for interpreting mass spectra have already been mentioned. Now that you are ready to test your understanding, most of these suggestions are presented again, but in somewhat abbreviated form. Bear in mind that the suggestions are just that—suggestions. As the student becomes more experienced in the area of mass spectrometry, he or she will develop his or her own unique interpretive style.

3.2 Some General Suggestions for Interpreting Mass Spectra

- Tabulate the m/z values of all major fragment and their relative abundances. Try to correlate as many of the fragments as possible with logical fragmentation patterns, paying close attention to the five or six most abundant fragments.
- Search the spectrum closely in the high mass region for the presence of a molecular ion. This is a valuable piece of information in the identification of an unknown compound.
- To every extent possible, an effort should be made to obtain as much knowledge as possible about the history of the sample.
- Remember, if M^+ has an odd molecular mass, the compound must contain an odd number of nitrogen atoms. If M^+ is even, the compound will contain either no nitrogens or an even number of nitrogens.
- If the molecular ion is absent, it may be beneficial to search the spectrum for fragments reflecting the loss of a neutral fragment from M^+. If such a fragment is verified, its usefulness is obvious in the indirect deduction of the molecular weight.
- Try to locate fragments whose combined masses give the molecular weight of the compound.
- If M^+ and $(M + 1)^+$ are measurable, use this information to determine the carbon number in conjunction with Beynon's

Table to approximate the probable empirical formula of the compound (Section 1.29).

- It should be emphasized that the $(M + H)^+$ fragment may be imposed upon $(M + 1)^+$ (the ^{13}C isotope fragment) resulting in only an approximation of the carbon number.
- Check the spectrum for an isotope cluster, and if present, compare its appearance with those in Figs. 1.16, 1.17, and 1.18.
- Look for broad, non-integral metastable fragments, and use their values to substantiate the structure of related fragments (Section 1.8). It should be noted, however, that some instruments are incapable of providing metastable fragments.
- Look for fragments at m/z 51, 63, 65, 76, 77, and 78, which suggest aromaticity.
- Remember, iodine and fluorine are monoisotopic, and their presence in a compound is often supported by fragments at m/z 127 (iodine) and 19 (fluorine).

3.3 Criteria for Finding the Molecular Ion

- If the compound is known, the molecular ion will have an m/z value equal to the molecular weight of the compound, assuming each element to be present as its most abundant isotope, and with atomic weights rounded to integer values.
- The nominal molecular weight, or m/z value for the molecular ion, will be an even number for any compound containing only C, H, O, S, Si, P, and the halogens.
- Fragment ions derived via homolytic bond breaking in these ions will have odd m/z values.
- Fragment ions derived from even mass molecular ions via expulsion of neutral components (e.g., H_2O, CO, ethylene, etc.) will have even m/z values.
- **The Nitrogen Rule:** A compound containing one or an odd number of nitrogen atoms (3, 5, 7, etc.) in addition to C, H, O, S, Si, P, and the halogens will have an odd molecular weight. An even number of nitrogen atoms in a compound will result in an even molecular weight.
- The molecular ion peak must have the highest m/z value of any significant (non isotopic or background) peak in the spectrum.

Corollary: The highest m/z value peak observed in the mass spectrum need not correspond to a molecular ion. For Cl and Br, the m/z value of lowest peak in molecular ion isotopic peak cluster is the molecular ion.

- The peak at the next lowest m/z value must not correspond to the loss of an impossible or improbable combination of atoms. Corollary: All non-isotopic peaks must correspond to the formulas that are subsets of the overall molecular formula.
- No fragment ion may contain a larger number of atoms of any particular element than the molecular ion.

3.4 Interpretation of a Mass Spectrum Using the National Institute of Standards and Technology (NIST) Mass Spectral Search Program

(It should be noted that NIST may be able to provide students with an inexpensive student version of the NIST Mass Spectral Search Software.)

(1) First, conduct the user spectrum search with the search mode set to Identity/Normal. If you find the value of InLib > 200 and the probability of the first hit > 0.9, you have almost certainly found the correct identification. If neither of these is the case, take note of the first 10 or so hits. Look for common structural elements.

(2) Repeat the user spectrum search, but use the search mode Similarity/Simple. You will not find values given for probability or InLib for this search. The similarity search is more likely to find compounds that have common substructural features than the identity search. Again, note the top 10 or so hits. You can save any hit in the hit list to the clipboard by double clicking on it.

(3) Click on your unknown spectra in the clipboard. This will make it the active spectrum. If you do not know the molecular weight of the unknown, from Tools on the menu bar, select MW

Estimation. This will give two different estimates of the molecular weight of the compound. Note these, and return to user spectrum search. Select the search mode Similarity/Hybrid. This search mode will make use of neutral loss logic and, therefore, needs to know the m/z of the molecular ion. The estimate given below the Similarity box is the same as the simple algorithm version of the MW estimator. You may want to search using each of the molecular weight values in turn. Enter a molecular weight into the box, and select search. Again, note the top 10 hits.

(4) At this point you should have a series of possible structures. If you have found certain common substructures in all of the searches, you may reasonably expect that these will be a part of your compound. You will have two independent estimates of the molecular weight of your compound. Using the substructures at hand and the molecular weight, you can attempt to produce candidate structures.

(5) It may be useful to use the neutral loss search by itself. If you do, note that the results of the search are very sensitive to the molecular weight entered. If you use this search, you may want to go beyond the molecular weight given by the MW estimation, and try a small range (2–3) on either side.

(6) It is better if you have independent information about the molecular weight such as from a chemical ionization mass spectra. Ideally, the results will be in accord with the other searches (meaning similar compounds in the top 10 hits) when the MW entered matches the sample. You should be aware that 10% to 15% of the spectra do not have molecular ions. The algorithms are very likely to fail here.

(7) If you have reason to suspect that the spectra may arise from a sample with significant impurities, then you may want to perform a reverse search. To do this, reset the search mode to identity and select Reverse, using the check box below. NOTE: Results found with reverse search can be very different and are affected more by incomplete library spectra than are other types of searching. Use caution.

(8) Spectral interpretation must be reconciled with all other information about the sample. If possible, check interpretation

by obtaining spectrum of assumed com pound under the same instrumental conditions.

3.5 Problems

(1) Briefly discuss the theory of mass spectrometry.
(2) What information can be deduced from a compound that has an even molecular ion?
(3) The mass spectrum of ethyl alcohol shows an intense ion fragment at m/z 31. What is the probable structure of the fragment?
(4) Draw structures for the two or three most abundant ions expected in the mass spectrum of 2,2-dimethylbutane.
(5) Explain why a monochloro compound shows an $(M + 2)^+$ ion.
(6) What structural conclusion can be drawn about a nitrogen containing compound with an odd molecular ion?
(7) List at least four fragment ions with m/z 28 values.
(8) What are the M and M + 2 ratios for an alkyl chloride containing one chlorine?
(9) List all possible isotopic contributions for dibromoethane.
(10) Explain the homologous fragments found in the mass spectrum of hexadecane shown in Fig. 3.1.

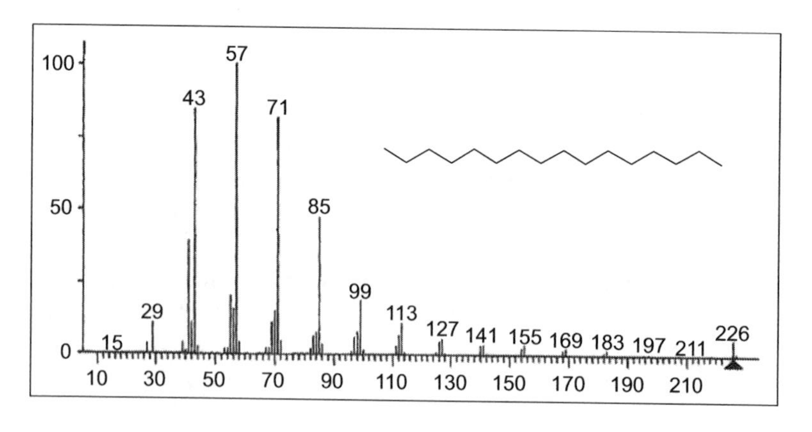

Figure 3.1 Hexadecane.

(11) Using the binomial expansion, $(a + b)^m$, calculate the approximate relative intensities of M, and M + 2 in bromoethane.

(12) The spectrum shown in Fig. 3.2 represents a straight-chain alkane. What is its structure?

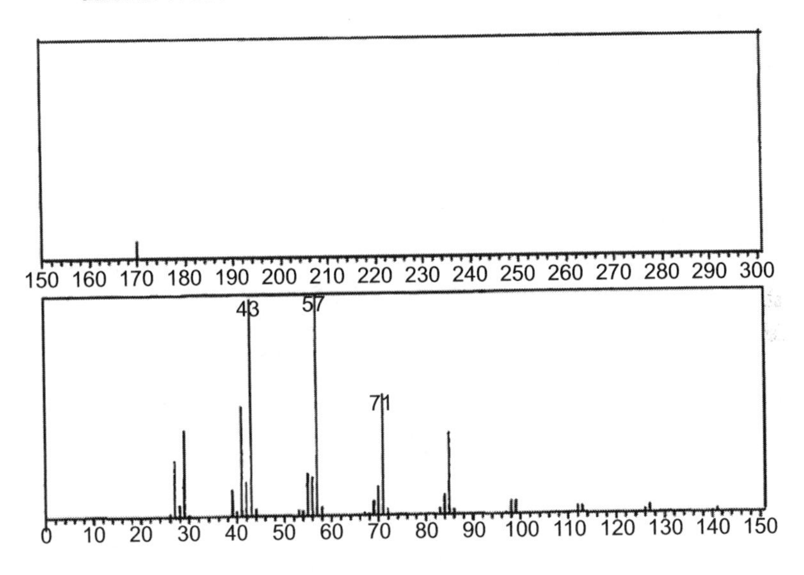

Figure 3.2

(13) Explain the large abundances of the fragments at m/z 57 and 85 in the mass spectrum of 2,2-dimethylpentane in Fig. 3.3.

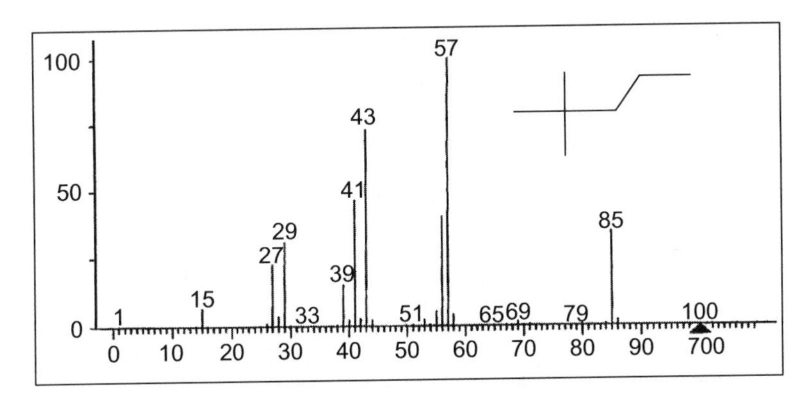

Figure 3.3 2,2-Dimethypentane.

(14) In the mass spectrum of propanal (Fig. 3.4), identify as many fragment ions as possible and explain the large abundance of the fragments at m/z 28 and 29.

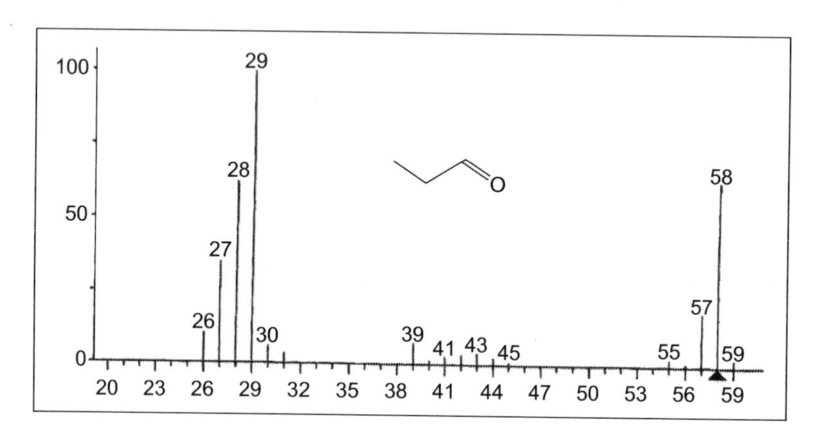

Figure 3.4 Propanal.

(15) In the mass spectrum of ethylbenzene (Fig. 3.5), identify as many fragment ions as possible and explain the large abundance of the fragment at m/z 91.

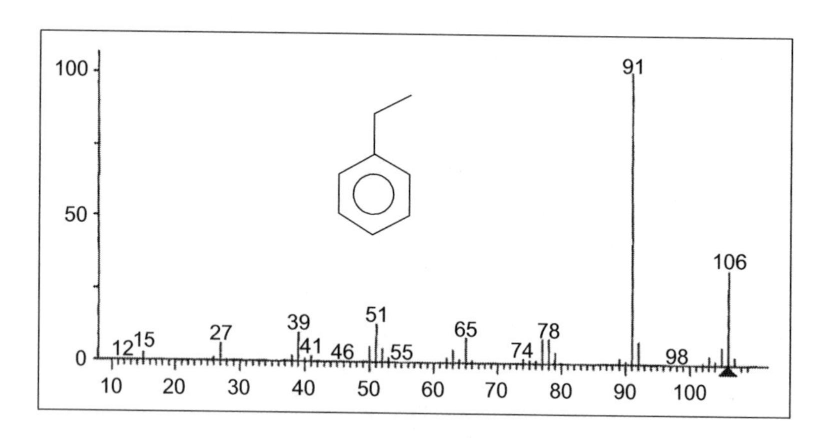

Figure 3.5 Ethylbenzene.

(16) The two spectra shown in Figs. 3.6 and 3.7 represent methyl-n-propyl ether and diethyl ether. Label each spectrum and give a rationale for your decision.

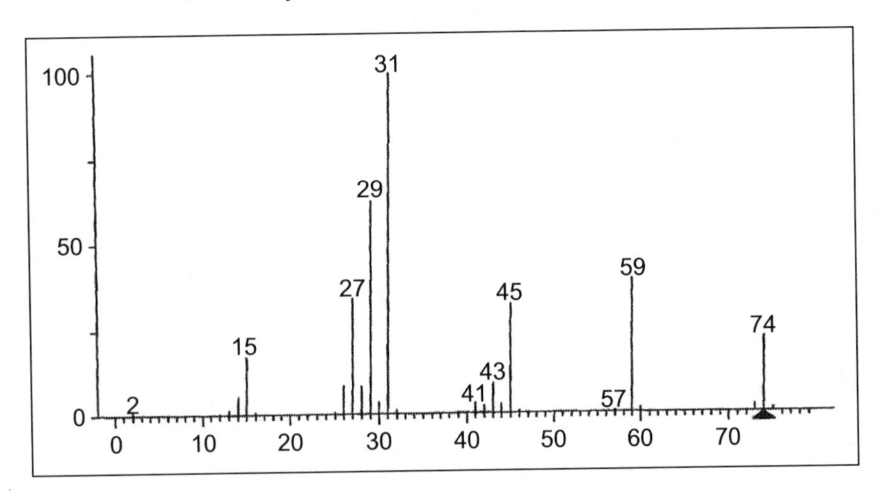

Figure 3.6

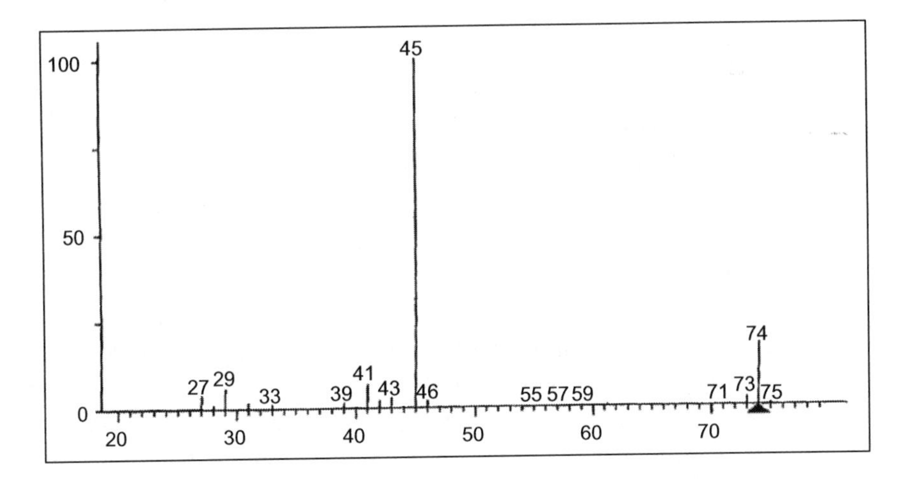

Figure 3.7

(17) Explain the large abundance of the fragment at m/z 43 in Fig. 3.8, and identify all other labeled fragments.

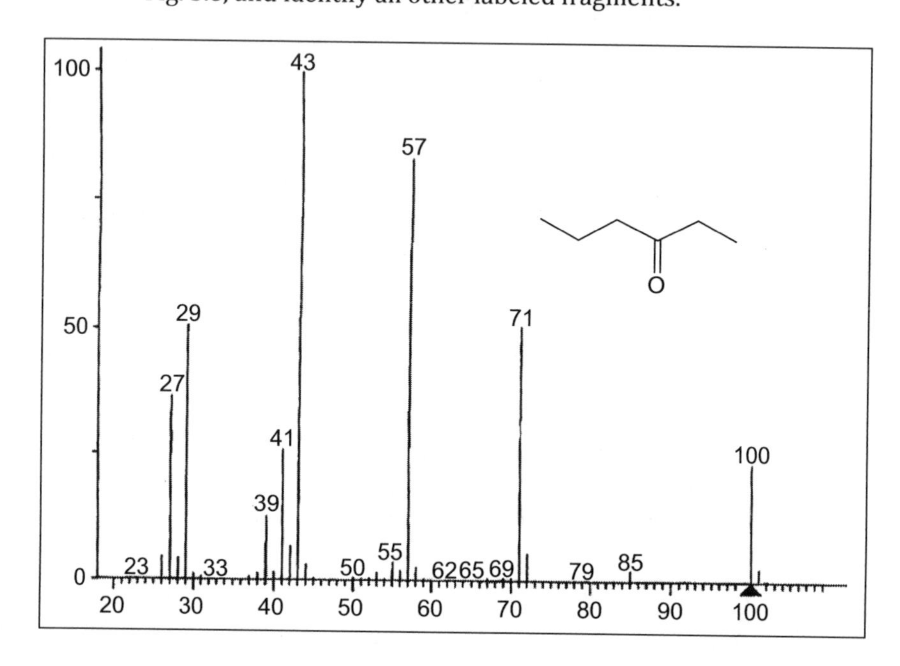

Figure 3.8

(18) Which of the compounds below would be expected to give the most abundant molecular ion? Explain.

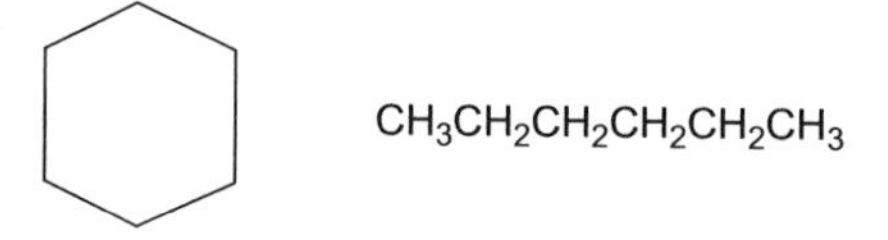

(19) Write a mechanism for the formation of the fragment at m/z 94 in the spectra of ethoxybenzene.

(20) Figure 3.9 represents a compound with the formula C_3H_8O. Suggest a structure for the compound and explain the formation of the fragment ions at m/z 45.

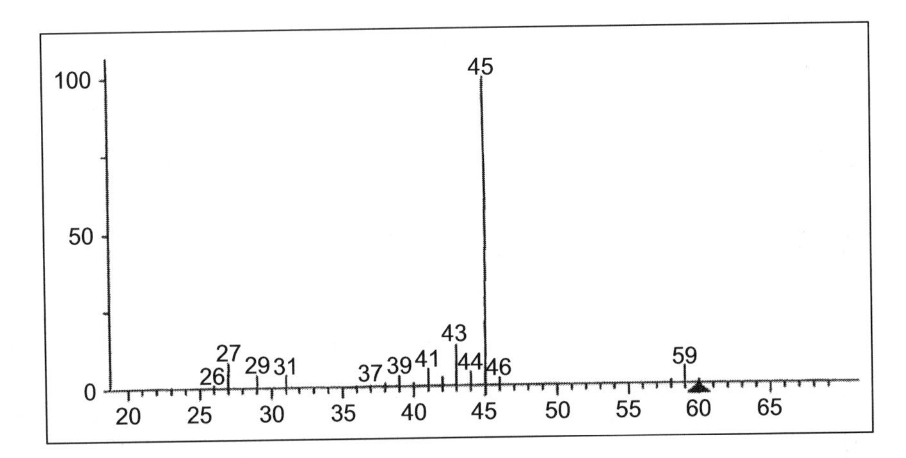

Figure 3.9

(21) Figure 3.10 represents a compound with the formula C_7H_7Br. Suggest a structure for the compound and explain the formation of the major ions.

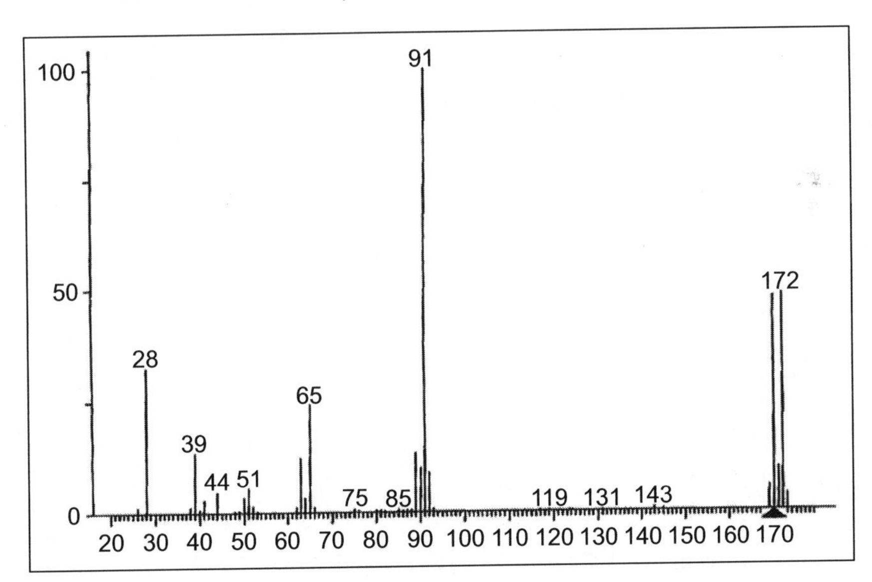

Figure 3.10

(22) Figure 3.11 represents a compound with the formula $C_8H_{10}O$. Suggest a structure for the compound and explain the formation of the labeled fragment ions.

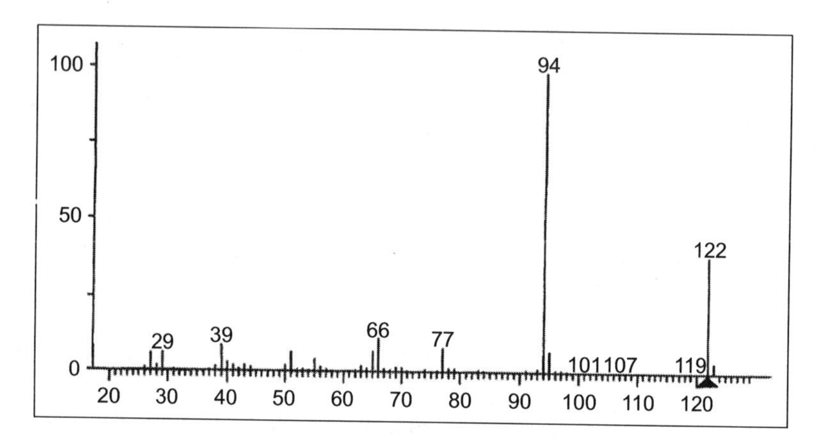

Figure 3.11

(23) Figure 3.12 represents a compound with the formula $C_4H_8O_2$. Suggest a structure for the compound and explain the formation of the labeled fragment ions.

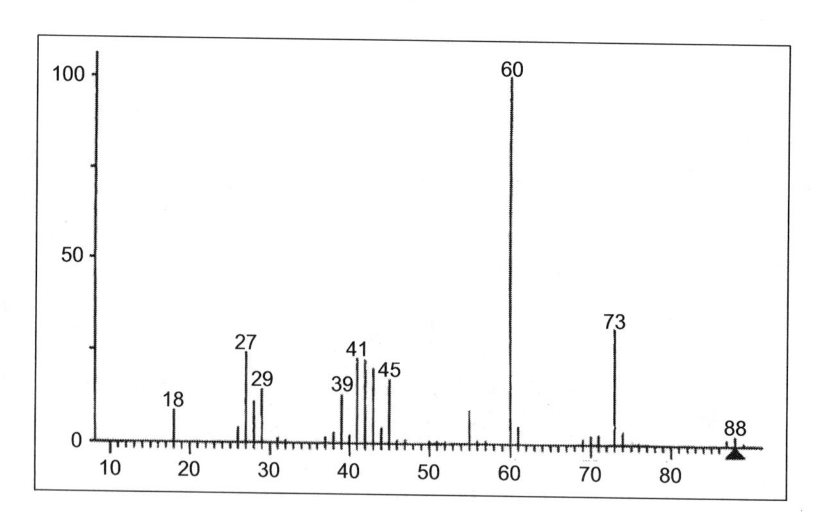

Figure 3.12

(24) Figure 3.13 represents an aromatic compound with the formula C_7H_8O. Suggest a structure for the compound and explain the formation of the labeled fragment ions.

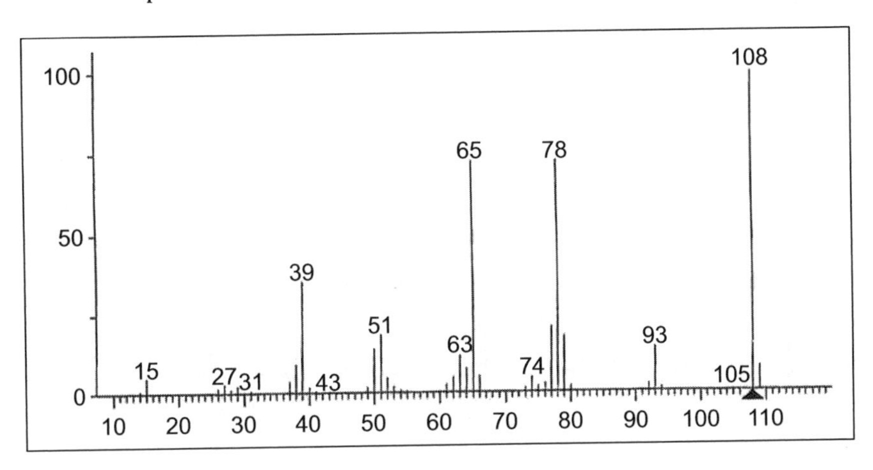

Figure 3.13

(25) In Fig. 3.14, draw structures for the fragment ions at m/z 105, 119, and 91.

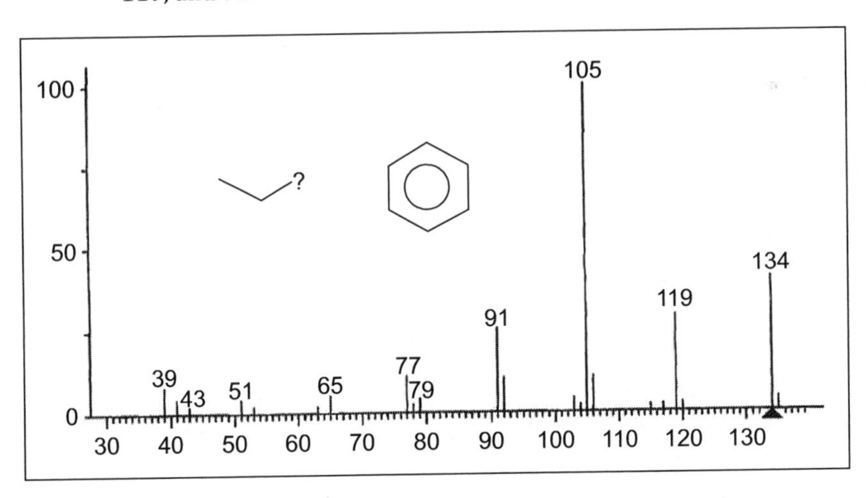

Figure 3.14 Diethylbenzene.

(26) In Fig. 3.15, draw structures for as many fragment ions as possible, including the major fragments at m/z 119 and 91.

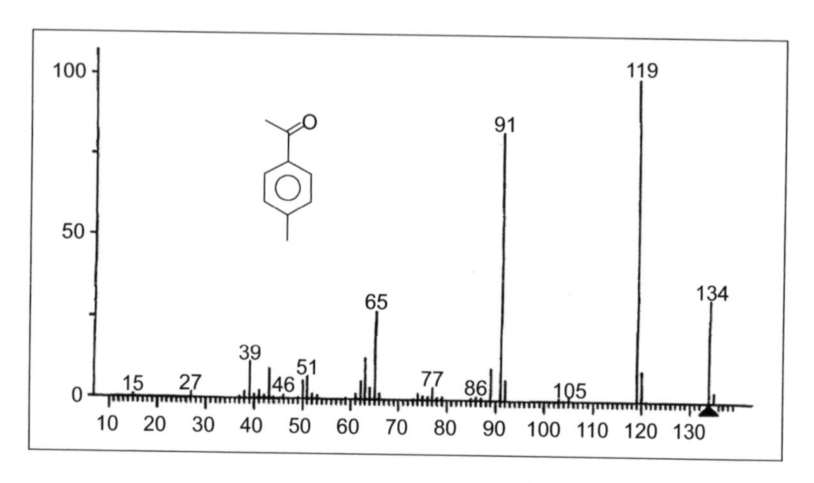

Figure 3.15 p-Methylacetophenone.

(27) In Fig. 3.16, draw structures for the fragment ions at m/z 59 and 101. Explain why no significant M^+ ion is present.

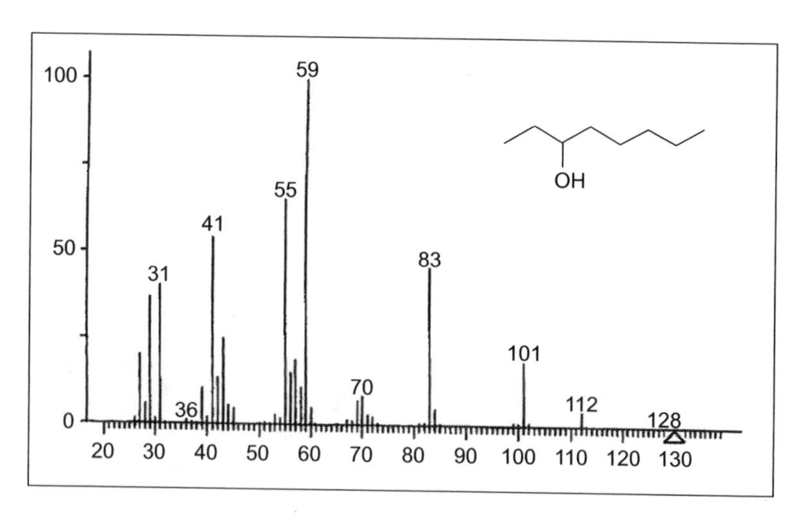

Figure 3.16 3-Octanol.

(28) In Fig. 3.17, draw structures for the fragment ions at m/z 42, 56, 70, 84, and 112.

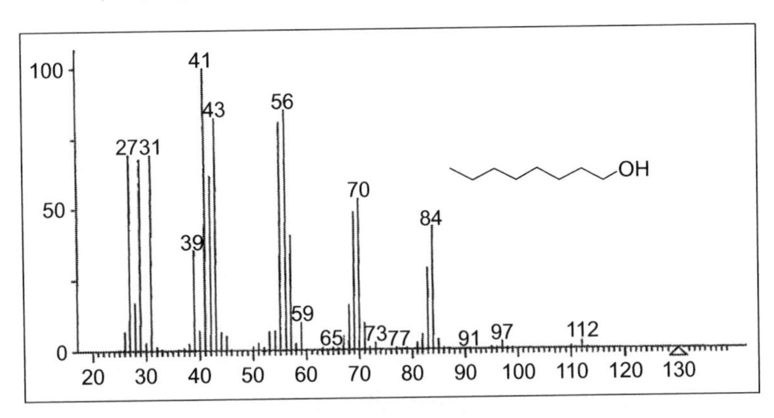

Figure 3.17 1-Octanol.

(29) In Fig. 3.18, draw structures for the fragment ions at m/z 57, 71, and 85. From the M^+ and $(M + 1)^+$ ions, calculate the approximate number of carbons present in the molecule and compare your value with the structure shown.

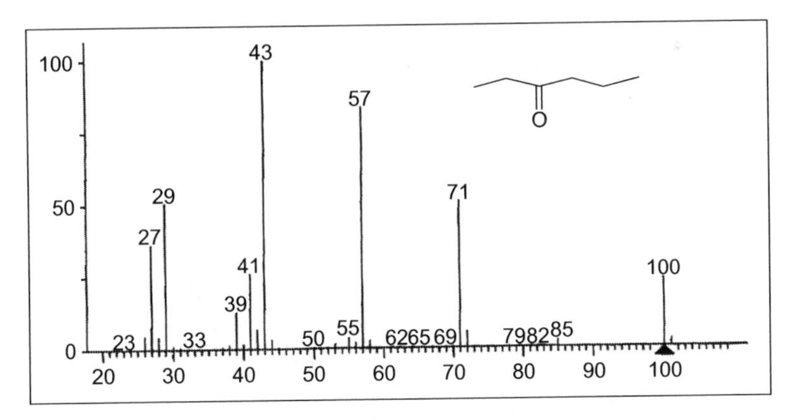

Figure 3.18 3-Hexanone.

(30) Identify the structures that correspond to the following spectra.

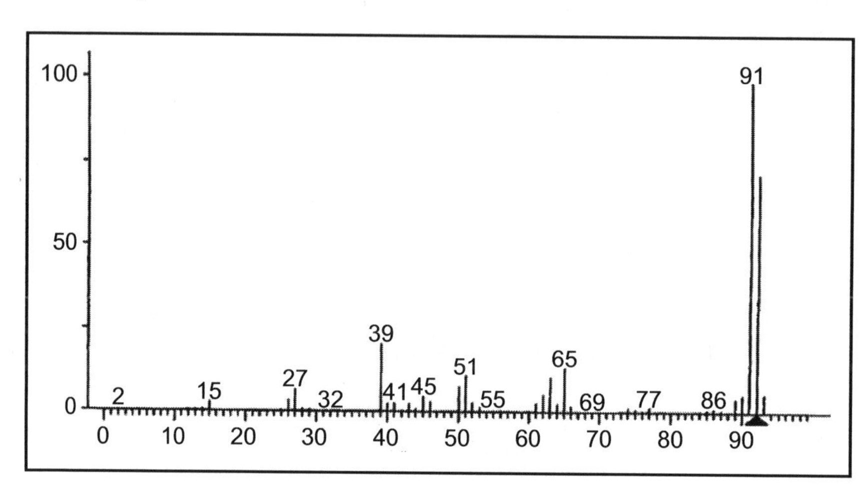

Figure 3.19 (C_7H_8).

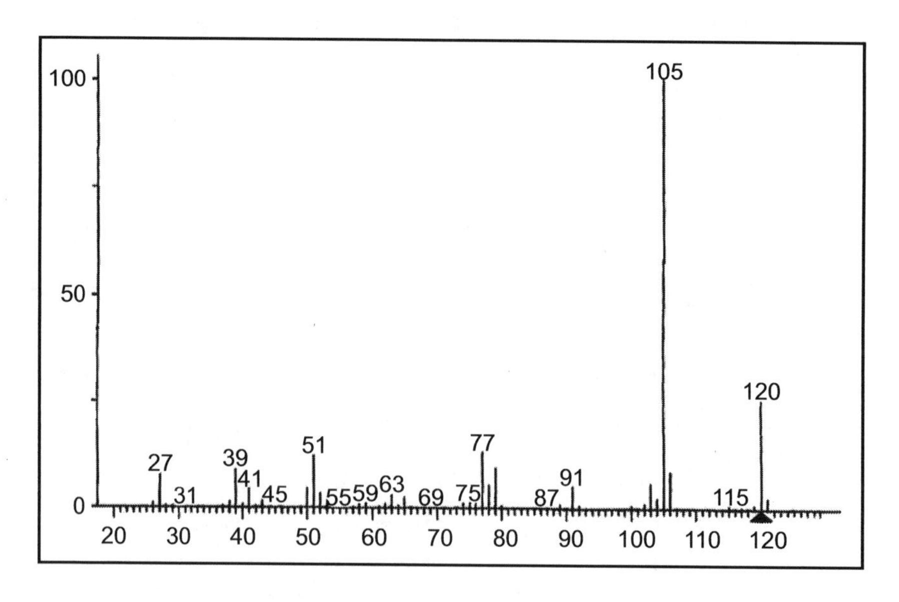

Figure 3.20 (C_9H_{12}).

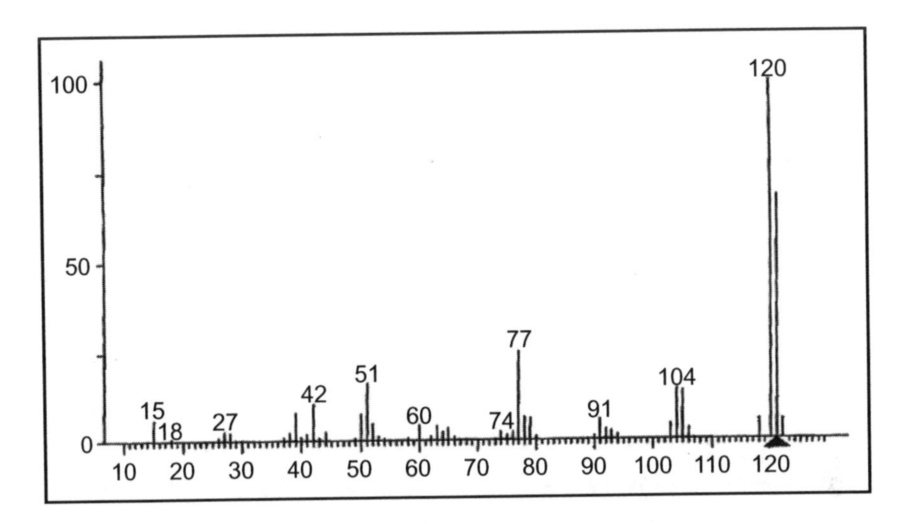

Figure 3.21 ($C_8H_{11}N$).

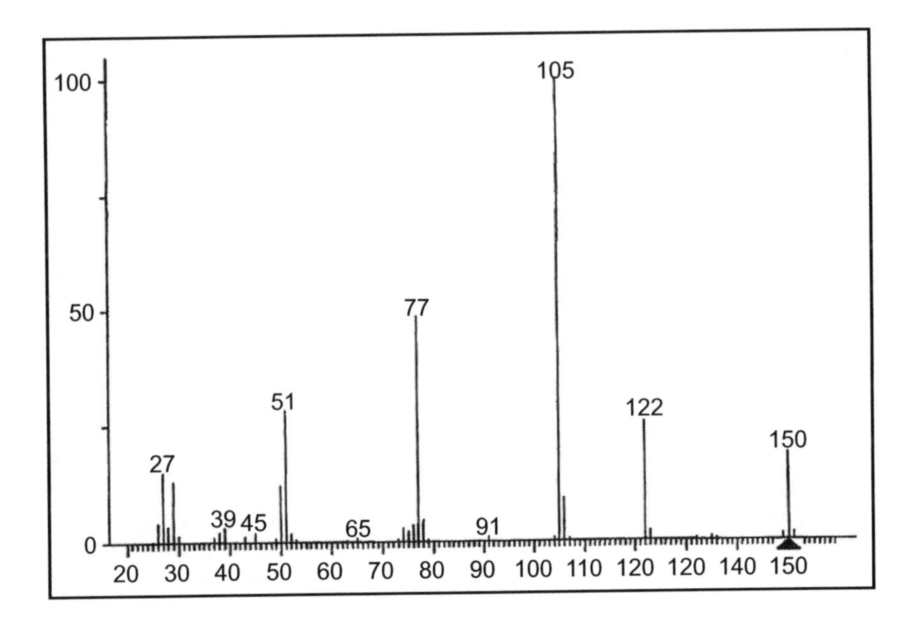

Figure 3.22 ($C_9H_{10}O_2$).

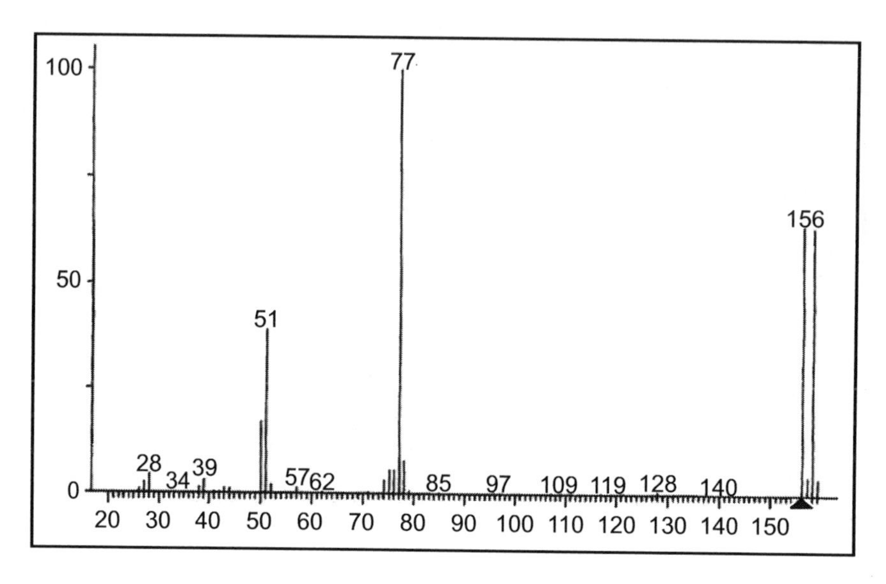

Figure 3.23 (C_6H_5Br).

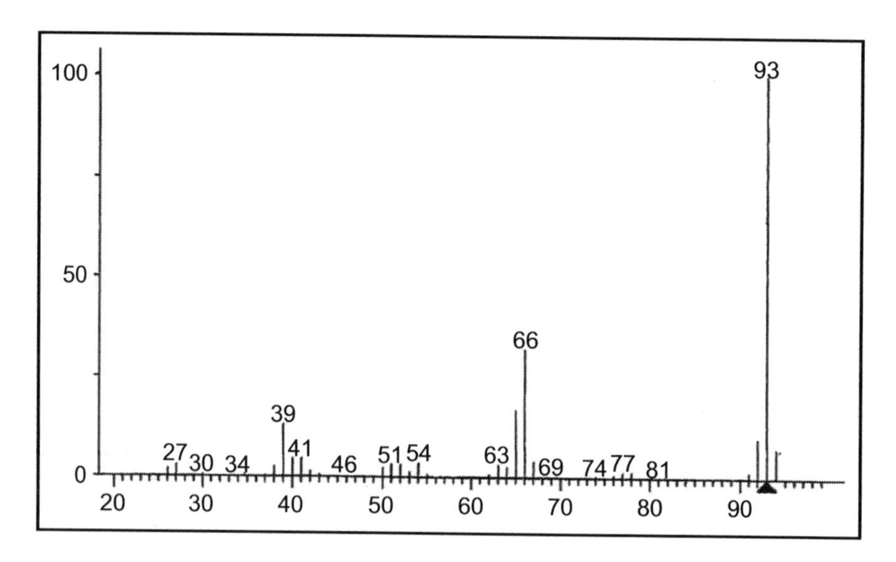

Figure 3.24 (C_6H_7N).

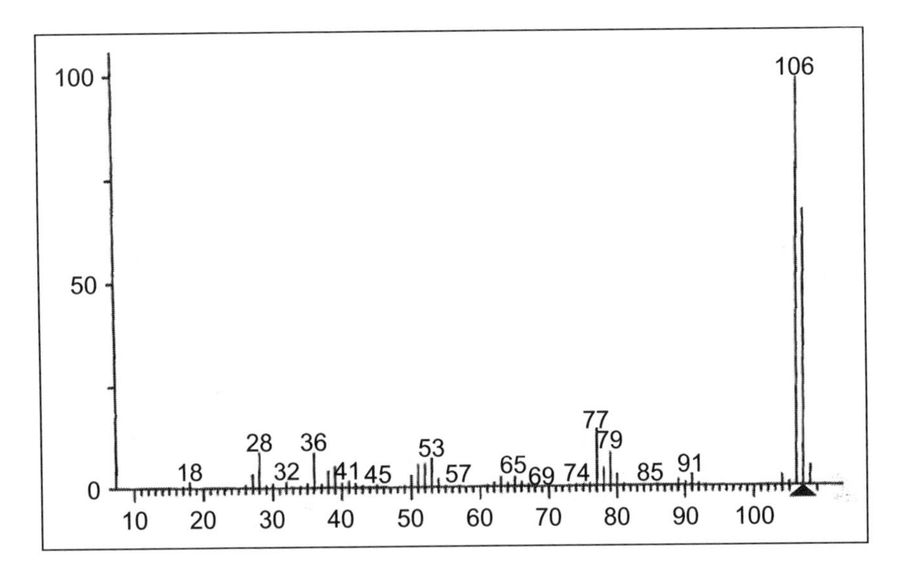

Figure 3.25 (C_7H_9N).

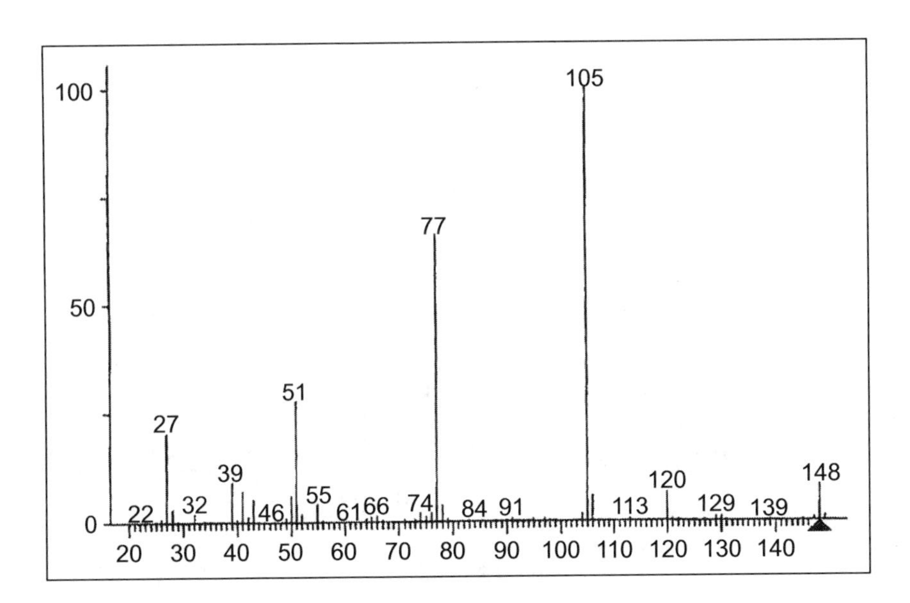

Figure 3.26 ($C_{10}H_{12}O$).

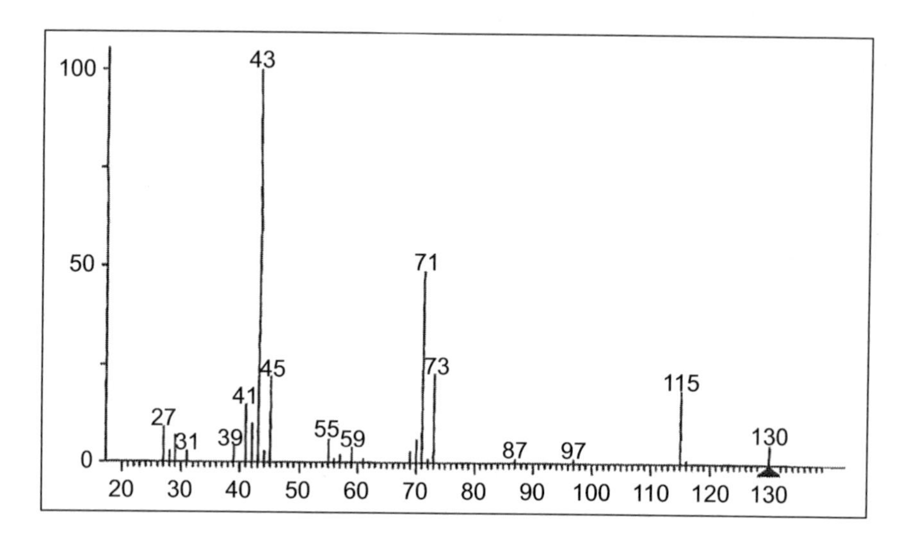

Figure 3.27 ($C_8H_{18}O$).

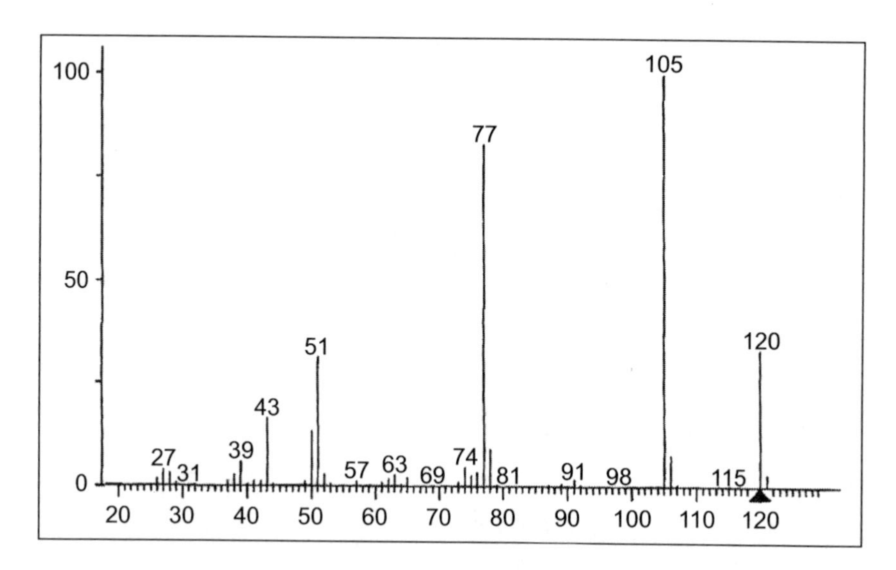

Figure 3.28 (C_8H_8O).

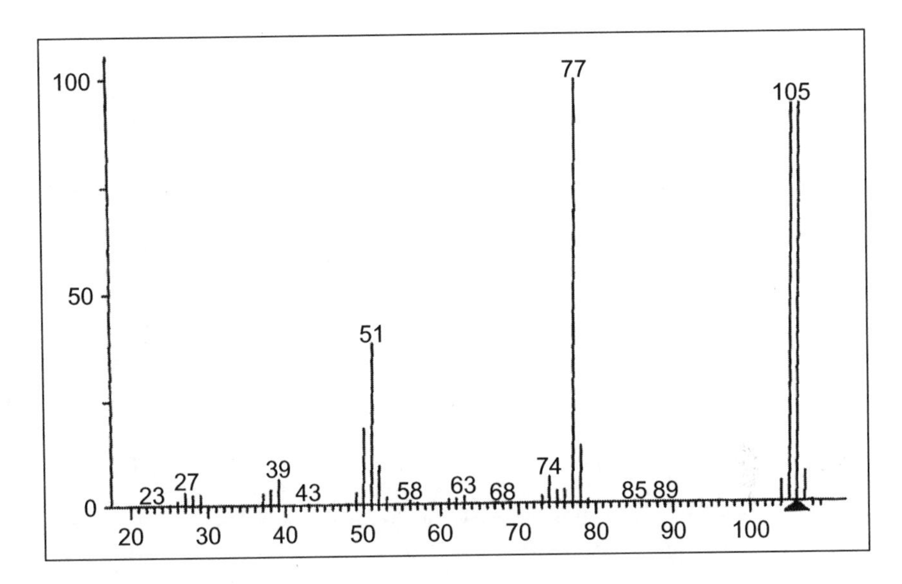

Figure 3.29 (C_7H_6O).

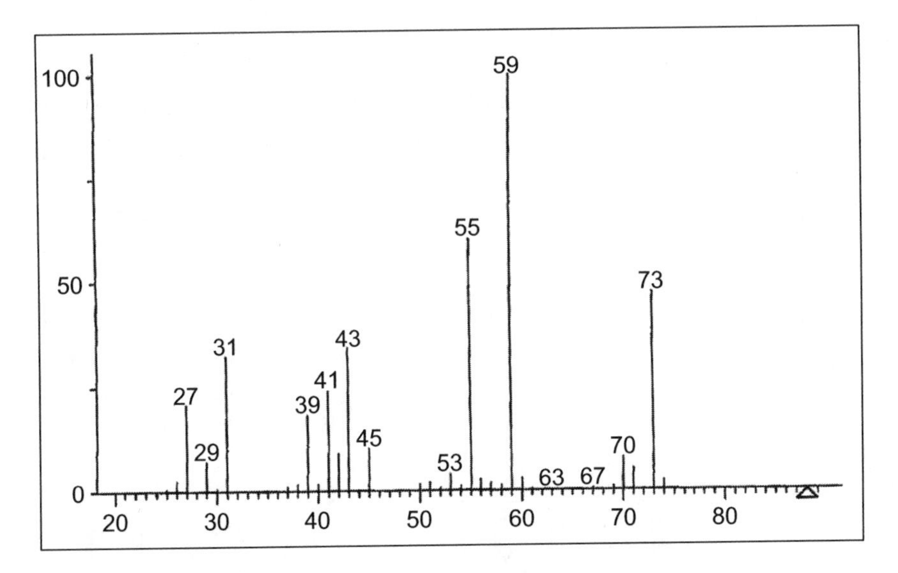

Figure 3.30 ($C_5H_{12}O$).

(31) What fragment or fragments in Fig. 3.31 suggest that the compound contains a phenyl ring? To what do you attribute the intensity of the m/z 119 fragment?

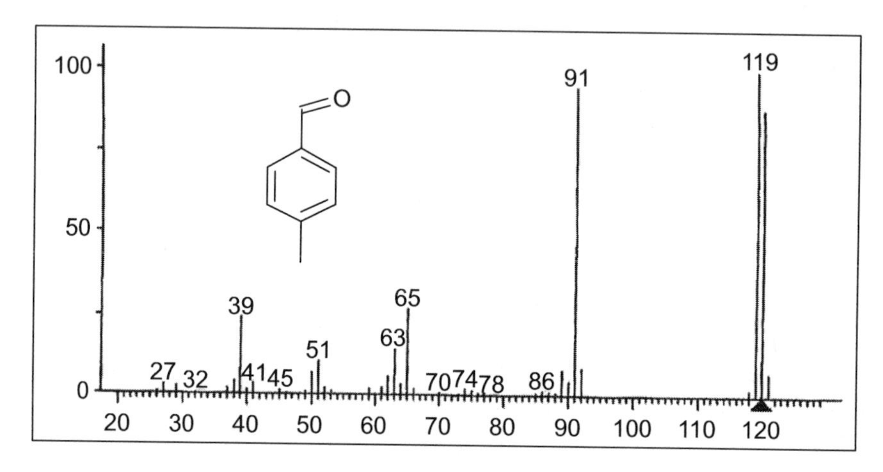

Figure 3.31

(32) Explain the significance of the $(M + 2)^+$ fragment in Fig. 3.32. What is the predicted $(M + 2)^+/M^+$ ratio? Explain. How does this value compare to the observed ratio?

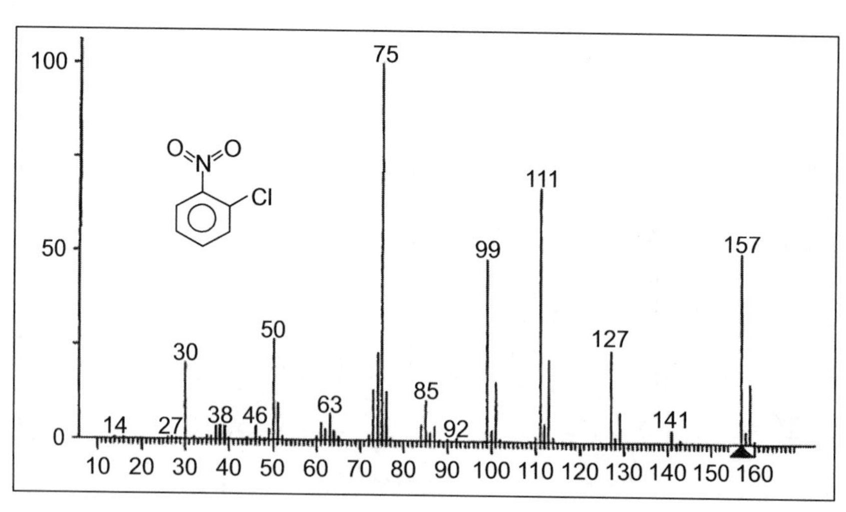

Figure 3.32

(33) Draw a probable structure of the base peak in Fig. 3.33. Discuss the origin of the fragment at m/z 69 and identify as many fragment as possible.

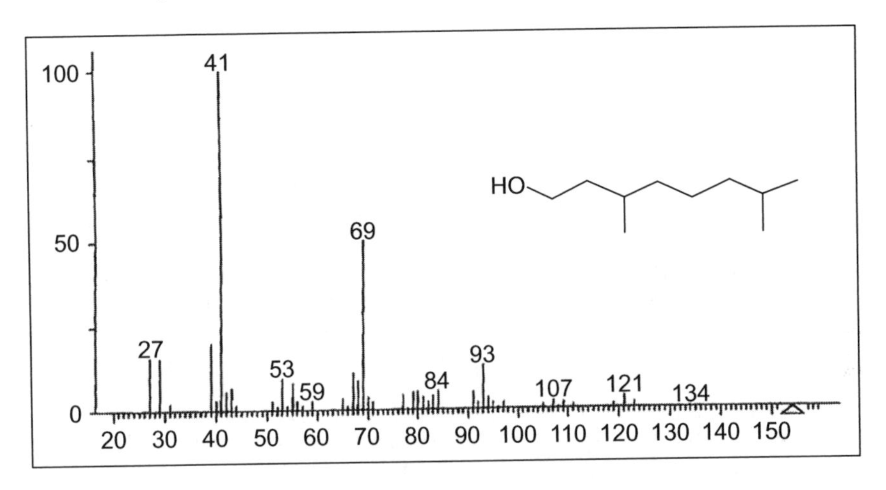

Figure 3.33

The questions that follow are based mainly on the NIST/EPA/NIH Mass Spectral Database.

(34) Based on the mass spectrum of acetophenone, predict the relative stability of the following fragment ions, and give an explanation for your decision.

$CH_3-C\equiv\overset{+}{O}$ m/z 43

$C_6H_5-C\equiv\overset{+}{O}$ m/z 105

(35) The base peaks in toluene, ethyl benzene, and benzyl chloride have identical m/z values. Explain its origin.

(36) The compound represented by the spectrum in Fig. 3.34 contains bromine and chlorine. Based on the isotopic patterns in Appendix A, determine the number or atoms of bromine and chlorine in the molecule.

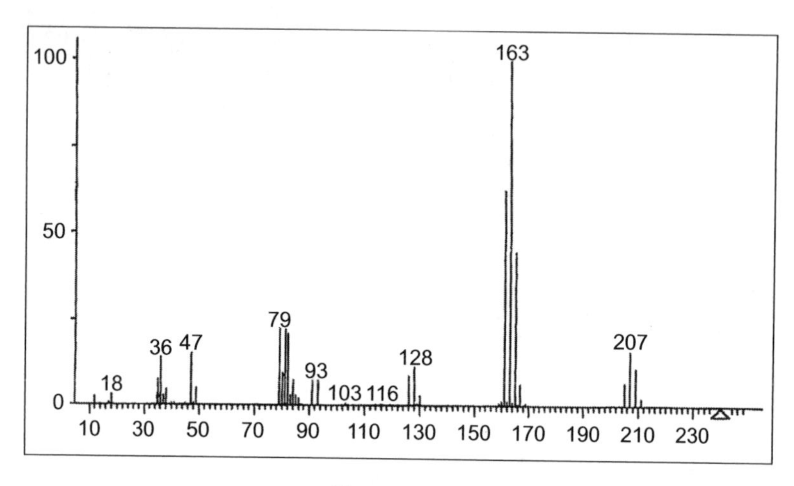

Figure 3.34

(37) The following tabulated data represent an unknown alkyl aryl hydrocarbon (MW 134). From the data, estimate the number of carbons in the unknown hydrocarbon.

	Mass	Abundance
M^+	134	237
$(M+1)^+$	135	25

(38) Based on the information in question 37 and the estimated carbon number, what is the structure of the unknown alkyl aryl hydrocarbon?

(39) Based on the spectrum in Fig. 3.35, what conclusion can be made about the structure of the compound relative to nitrogen?

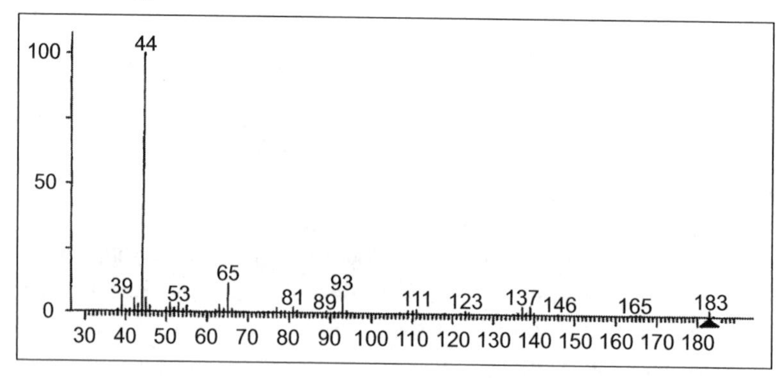

Figure 3.35

(40) Based on the M^+ and $(M + 2)^+$ ion in Fig. 3.36, what can you say about the structure of the unknown sample material with regards to halogens?

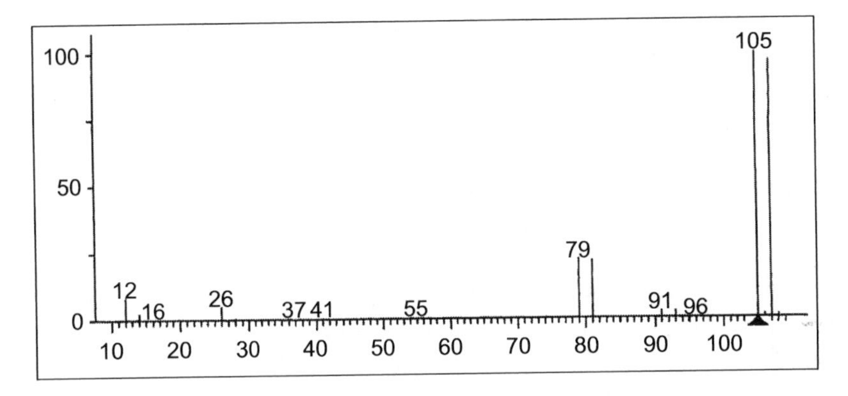

Figure 3.36

(41) The following compound has at least 99 different names. Why are there so many names for this compound?

H_2N — S(=O)(=O) — NH_2

(42) The base peak for butanal, shown below, is located at m/z 44. What is the structure of this fragment?

O

(43) Using the NIST/EPA/NIH Mass Spectra Database, determine the structures associated with the two mass spectra shown in Figs. 3.37 and 3.38.

Procedure for solving question 43 (also see Section 3.6.2)

(1) Determine the base peak in each spectrum and give each base peak an abundance range of 90 to 100.
(2) Determine the masses of the next two or three most abundant fragments in each spectrum, and give them an estimated abundance range based on the base peak. For instance, if one of the fragments has an abundance of about 50% of the base peak give it an abundance range of 50 to 60 or 45 to 55.
(3) Place this information in the "any peak" software program (under Search on the NIST menu).
(4) When you are finished, your search data should be similar to what is shown in the following two tables. (*Note*: The number of hits are generated by the software.)

Figure 3.37

m/z	from	to	peaks	hits
197	90	100	1	410
225	80	90	2	1

Figure 3.38

m/z	from	to	peaks	hits
108	90	100	1	774
109	85	95	2	6
241	70	80	3	1

As noted from the above data, the search related to Fig. 3.37 gave one hit after only two sets of data had been entered into the "any peak" program, and this one hit gave the correct structure, since the two mass spectra (known and unknown) were identical. In Fig. 3.38, after entering three sets of data, one hit was obtained and this hit gave a spectrum that was identical to the unknown spectrum of Fig. 3.38. The structures of the two unknowns are shown following Figs. 3.37 and 3.38.

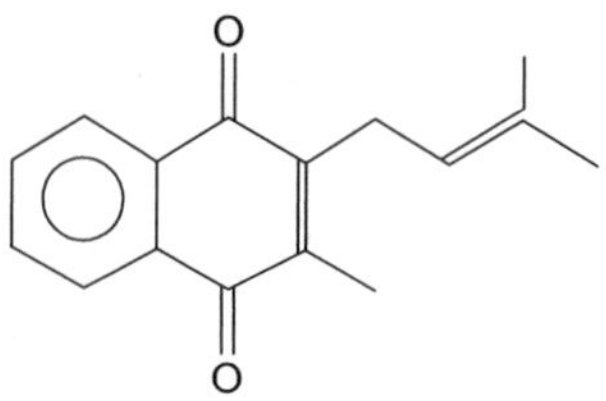

Structure Related to figure 3.37

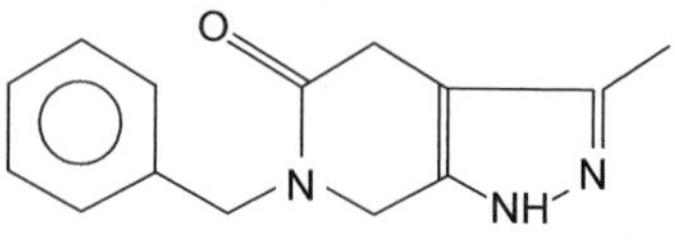

Structure Related to figure 3.38

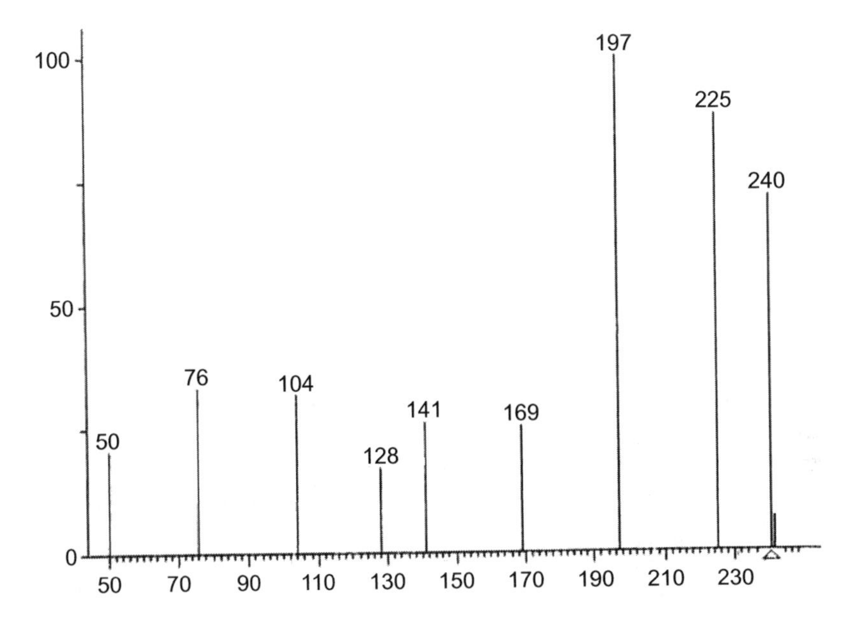

Figure 3.37

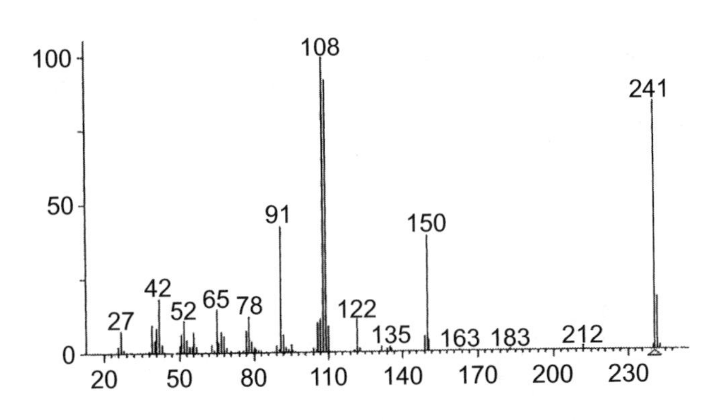

Figure 3.38

In situations where the unknown mass spectrum is in the computer in the proper format, it can be searched against a mass spectra database such as Wiley and NIST. However, if all you have is a printed copy of the mass spectra, then the solution is the NIST "any peak" software.

(44) Using the NIST software, match the spectra that follow with the appropriate structures.

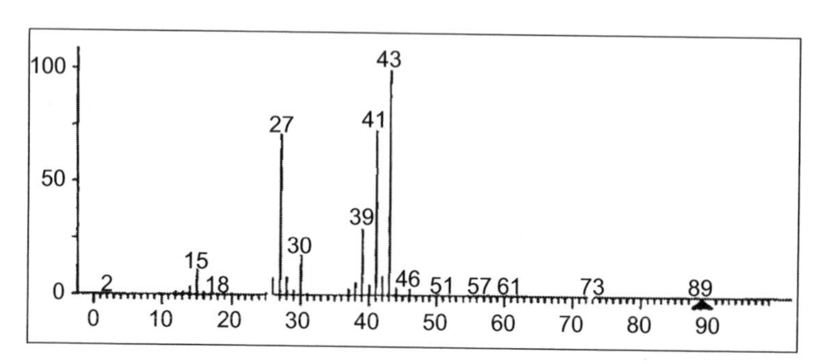

Figure 3.39

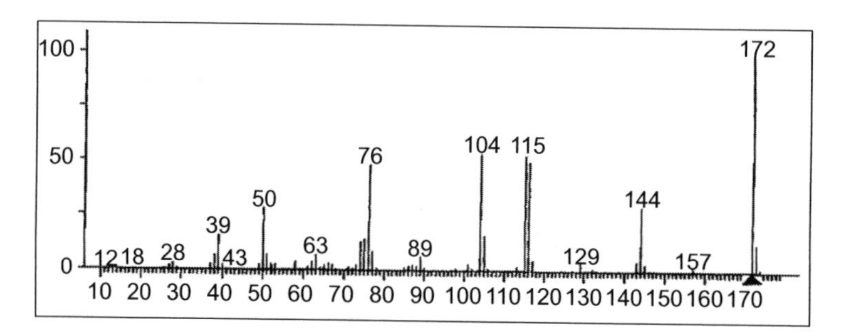

Figure 3.40

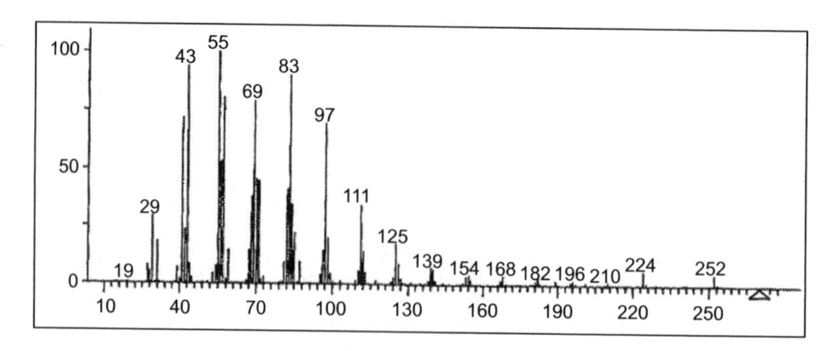

Figure 3.41

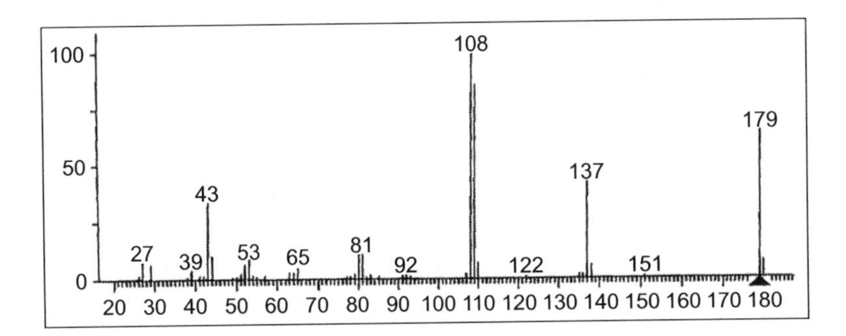

Figure 3.42

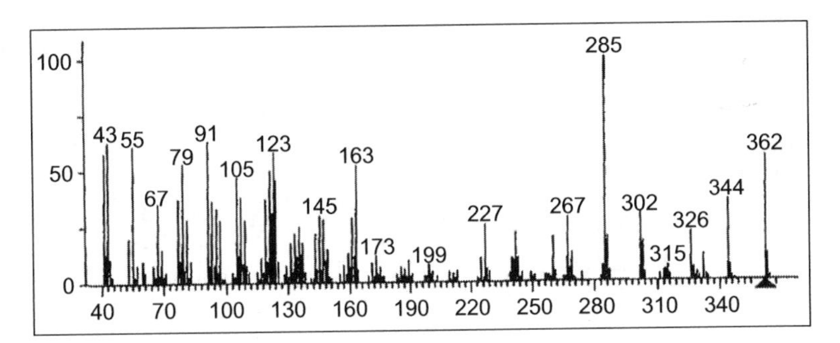

Figure 3.43

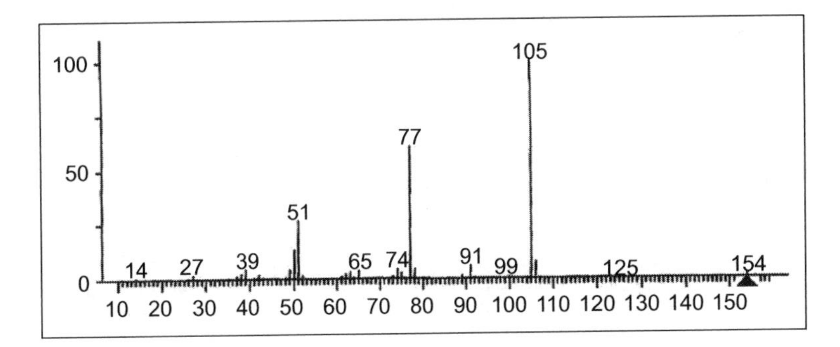

Figure 3.44

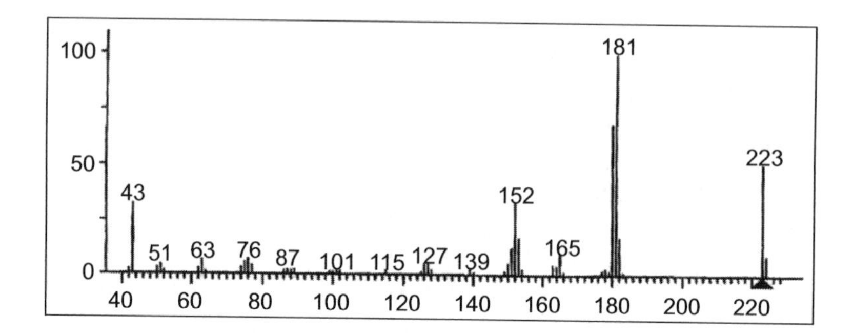

Figure 3.45

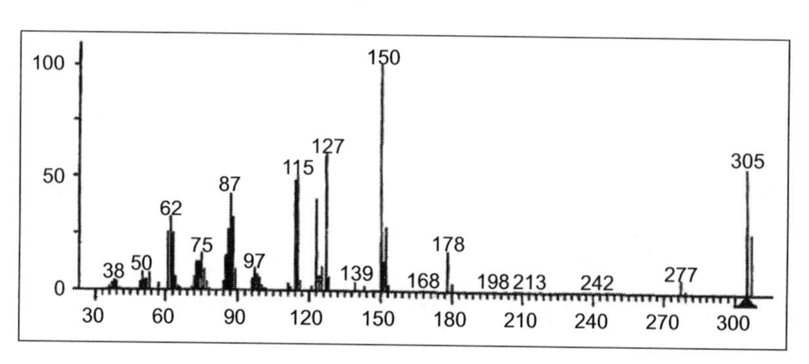

Figure 3.46

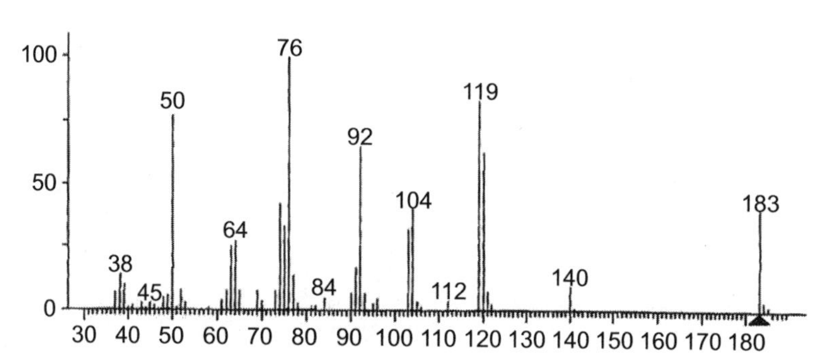

Figure 3.47

(45) Consider the spectra in question 44 (Figs. 3.38–3.47) and answer the following questions:

(a) Which spectra represent the compounds that definitely contain nitrogen? Explain.

(b) Identify all spectra containing bromine and/or chlorine and determine the number of bromine and/or chlorine. Explain.

(c) Identify all spectra that definitely correspond to aromatic compounds. Explain.

(d) Select all spectra that show the fragment ion below. Explain.

$C_6H_5-C\equiv\overset{+}{O}$

(e) Which spectrum shows the "hydrocarbon cluster"?

(f) Which spectrum (or spectra) shows the presence of iodine?

(46) Show a plausible mechanism for the formation of the base peak at m/z 44 in the spectrum of butanal.

(47) Consider fragment ions a and b, shown in the following diagram. From a study of the mass spectrum of 2,2′-dihydroxy-4-methoxybenzophenone (Fig. 3.48), determine which of the fragment ions, a or b, has the greatest relative stability.

CH_3O–(2-OH)C_6H_3–C(=O)–C_6H_4(2-OH) → CH_3O–(OH)C_6H_3–$C\equiv\overset{+}{O}$ (a) + (OH)C_6H_4–$C\equiv\overset{+}{O}$ (b)

2,2'-dihydroxy-4-methoxybenzophenone

(48) The fragmentation of benzyl benzoate results in the formation of fragment ions a and b shown in the following diagram. From a study of the mass spectrum of benzyl benzoate (Fig. 3.49), determine the relative stability of the two ions.

C_6H_5–C(=O)–OCH_2–C_6H_5 ⟶ C_6H_5–$C\equiv\overset{+}{O}$ (a) + C_6H_5–$\overset{+}{C}H_2$ (b)

(49) The base peak for 3-chloro-1-propene (Fig. 3.50) is located at m/z 41. What is the structure of this fragment ion? Explain its relative stability.

(50) Explain why the molecular ion (M^+) is much more abundant in cyclooctane than in octane.

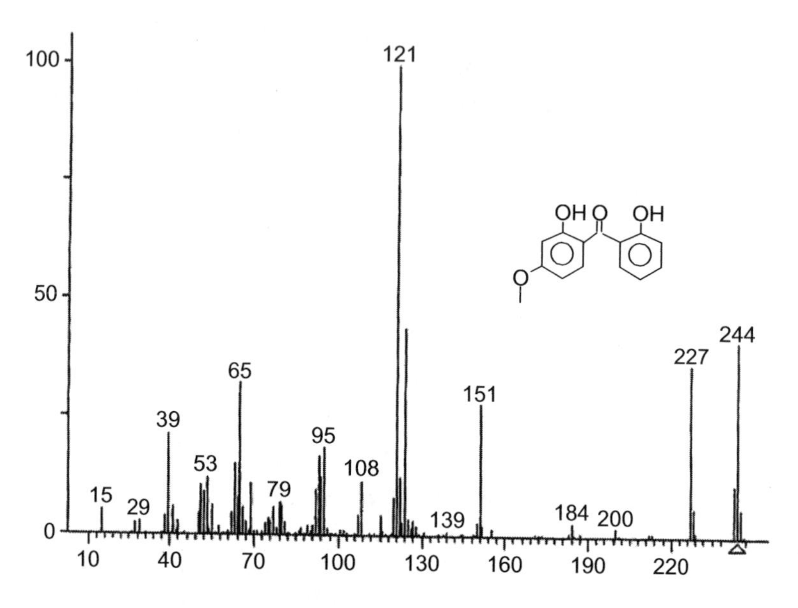

Figure 3.48

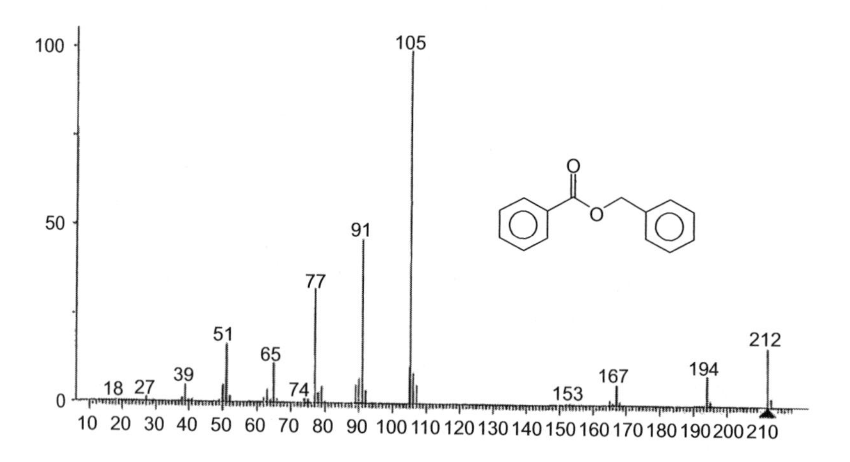

Figure 3.49

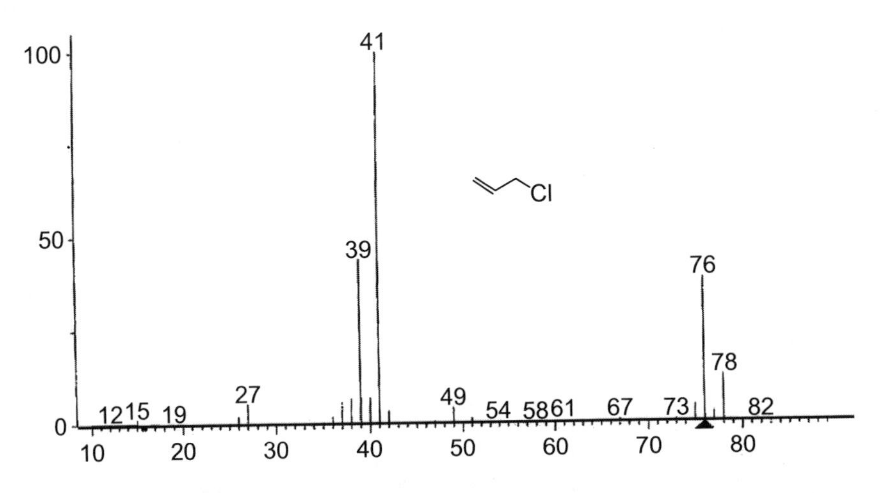

Figure 3.50

(51) In the following mass spectrum of acetophenone, what are the fragments at $m/z = 43$ and 105?

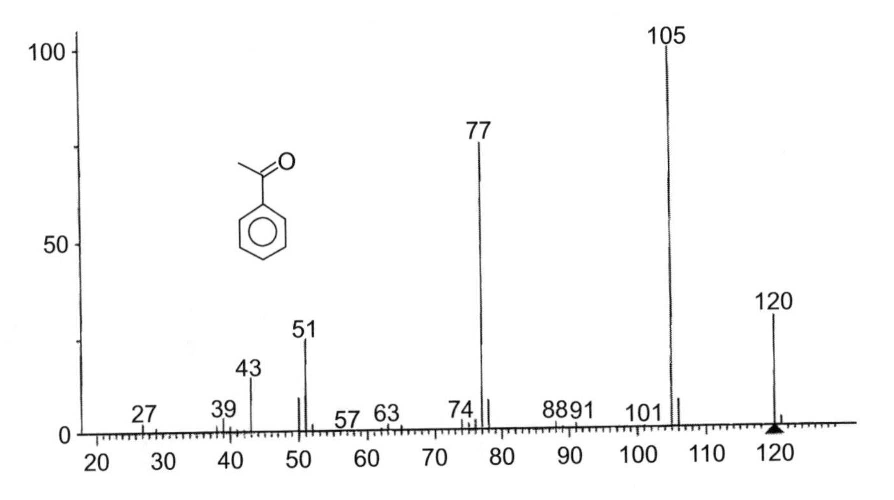

(52) In the following spectrum of n-propylbenzene, what are the fragments at $m/z = 91, 77$ and 52?

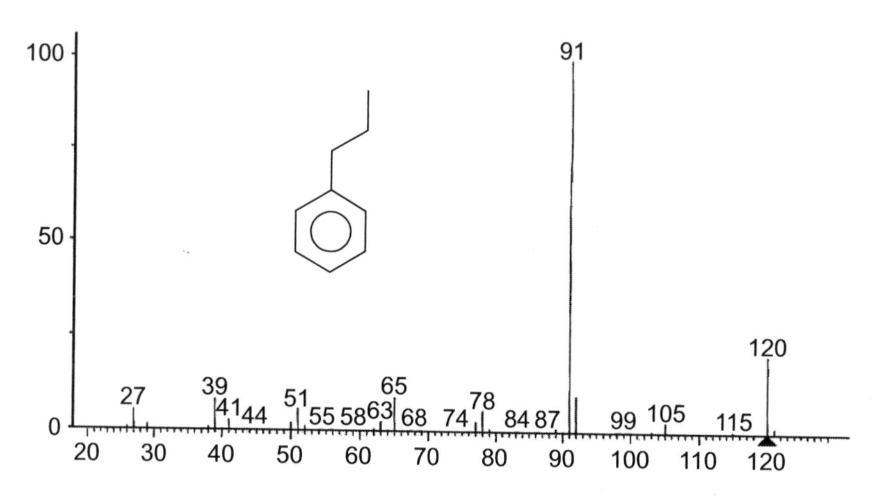

(53) The following mass spectrum is represented by one of the accompanying structures. Based on the fragment ions, determine which structure is correct.

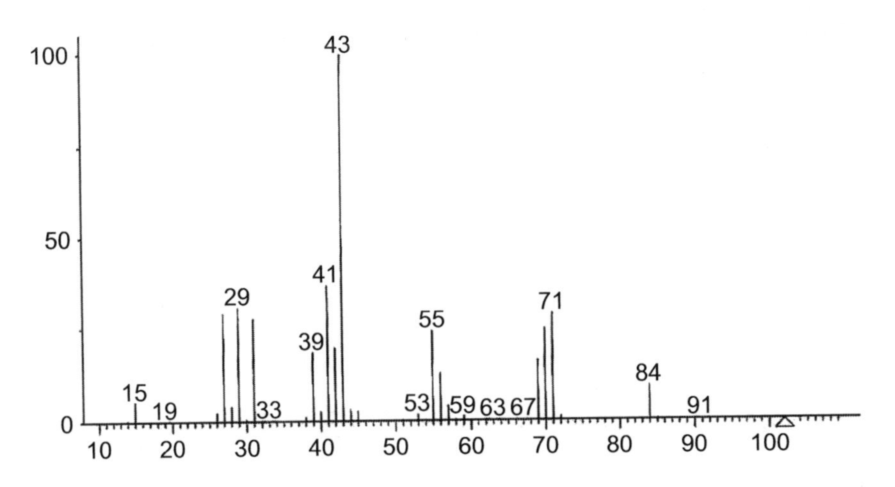

(54) Does the compound represented by the following mass spectrum contain halogen? If so, what is the halogen and how many halogens?

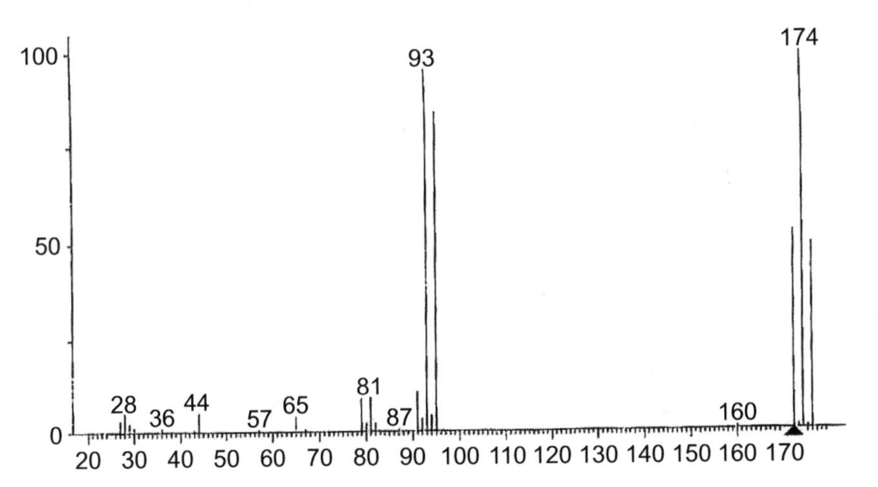

(55) Does the compound represented by the following mass spectrum contain halogen? Explain. If so, what is the halogen and how many halogens? Is nitrogen present in the compound? Explain.

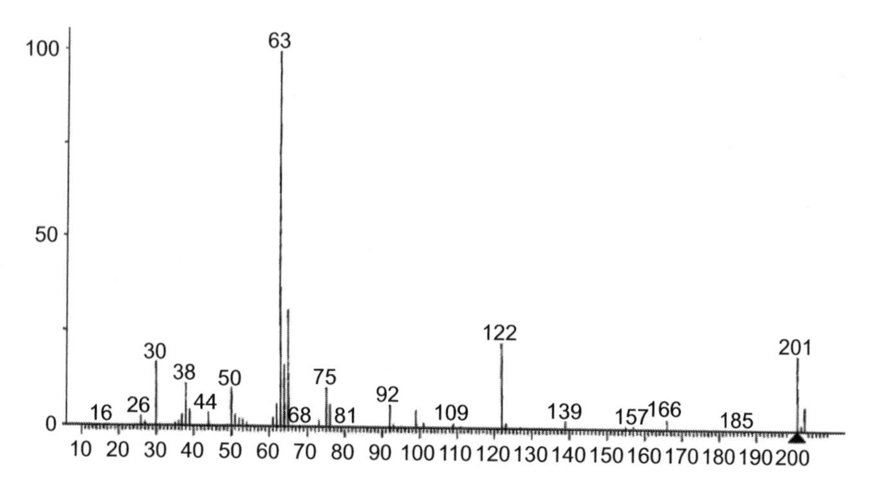

(56) In the following spectrum of a hydrocarbon, what are the structures of the fragments at $m/z = 29, 43, 57, 71, 85$, and 98?

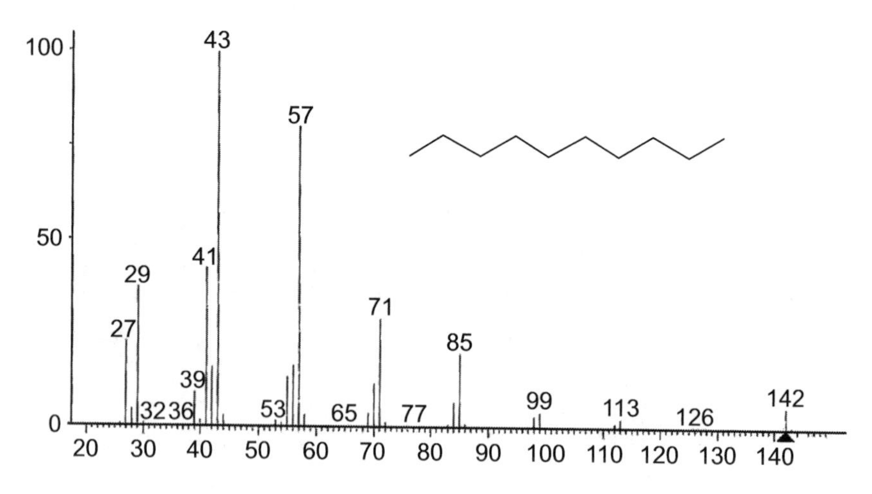

(57) Consider the following mass spectrum and estimate the relative stability of the benzoyl carbocation ($m/z = 105$) vs. the tropylinium ion ($m/z = 91$).

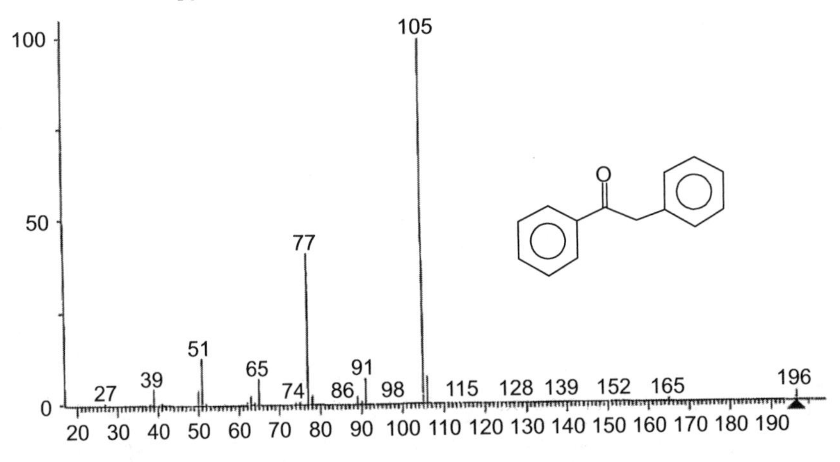

(58) In the following spectrum, using relative abundances, estimate the relative stability of the two carbocations shown above the spectrum:

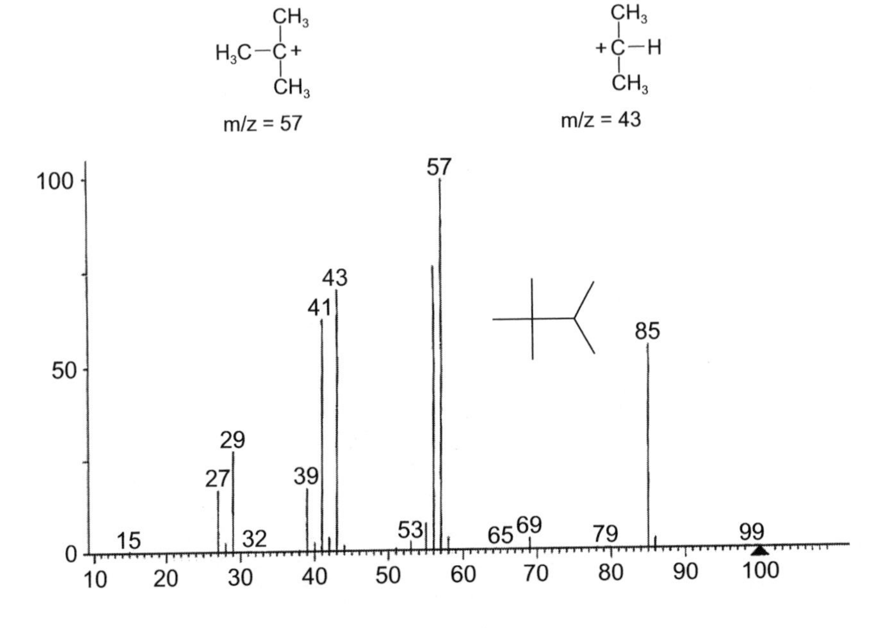

(59) The compound represented by the following spectrum contains halogen. What is the halogen, and how many? What is the probable structure of the fragment at $m/z = 91$? What is the probable structure of the compound?

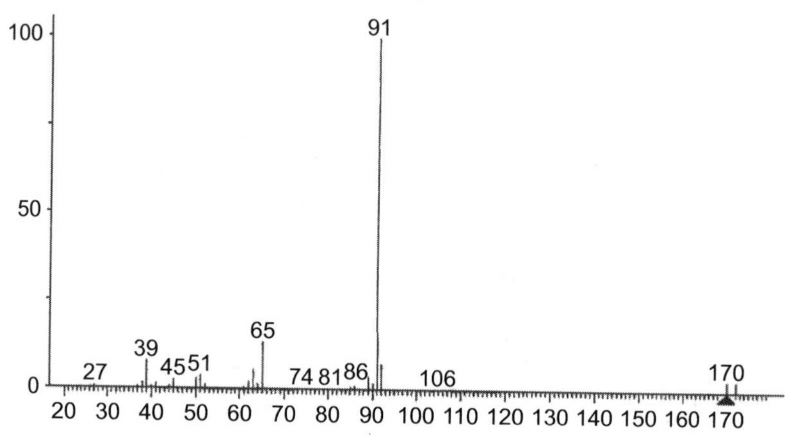

(60) The compound represented by the following mass spectrum contains nitrogen. What can you say relative to the number of nitrogens in the compound? What is the probably structure of the compound? Explain.

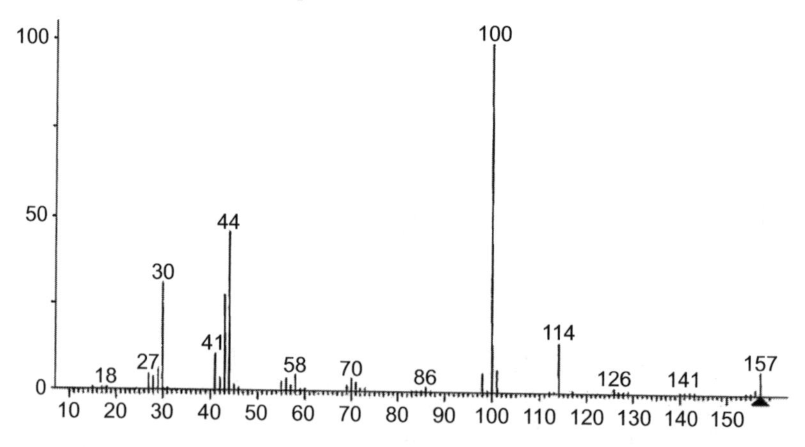

(61) The following spectrum represents a non-branched hydrocarbon. What is the structure of the hydrocarbon? Explain.

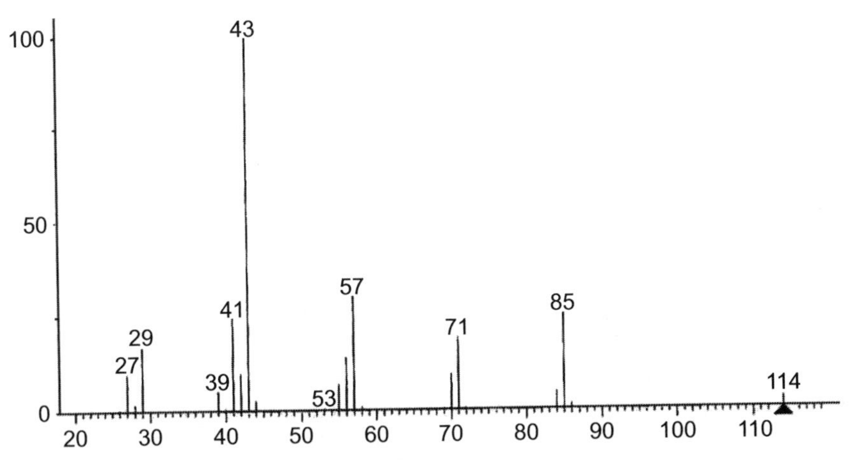

(62) The following structures (1–36) represent the 36 mass spectra that follow. Assign each structure to its correct mass spectrum and draw the structures of the four most abundant fragments in each spectrum.

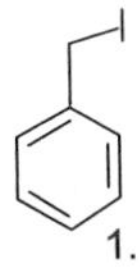

1.  2. 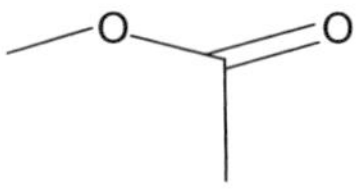 3. 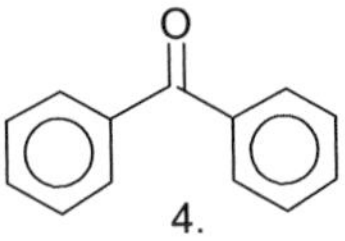 4.

NH

5. 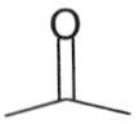 6. 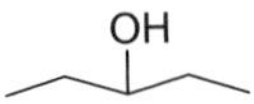 7. 8.

OH

$H_3C{-}C{\equiv}C{-}CH_3$

Br OH

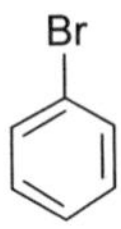

9. 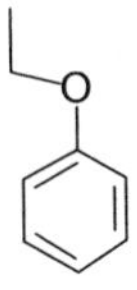 10. 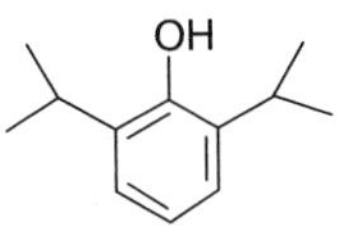 11. 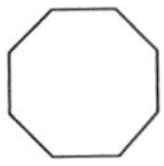 12.

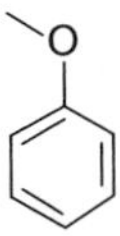

13.  14.  15. 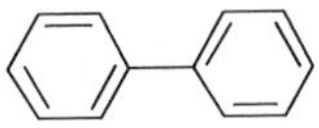 16.

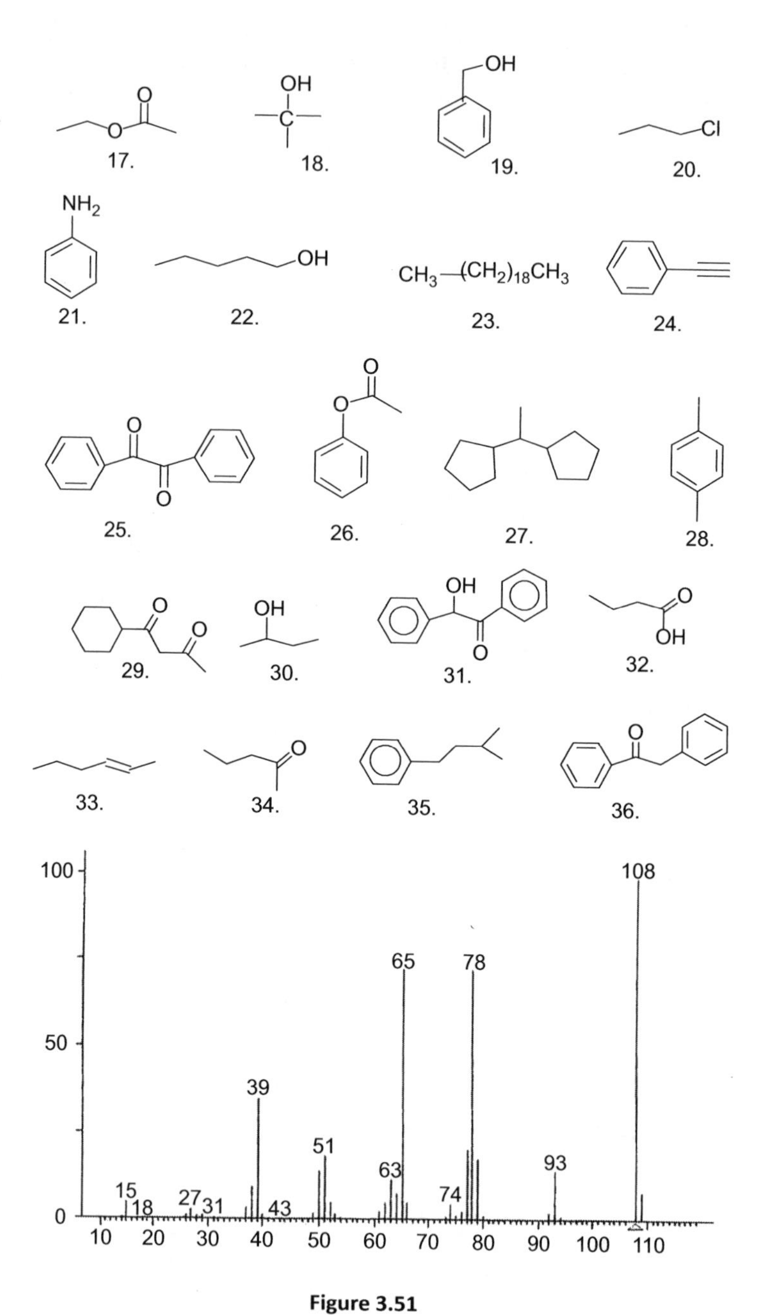
17.
OH
C
18.
OH
19.
Cl
20.
NH2
21.
OH
22.
CH3—(CH2)18CH3
23.
24.
O
O
25.
O
O
26.
27.
28.
O
O
29.
OH
30.
OH
O
31.
O
OH
32.
33.
O
34.
35.
O
36.
100
50
0
108
65
78
39
51
63
93
74
15
18
27
31
43
10
20
30
40
50
60
70
80
90
100
110

Figure 3.51

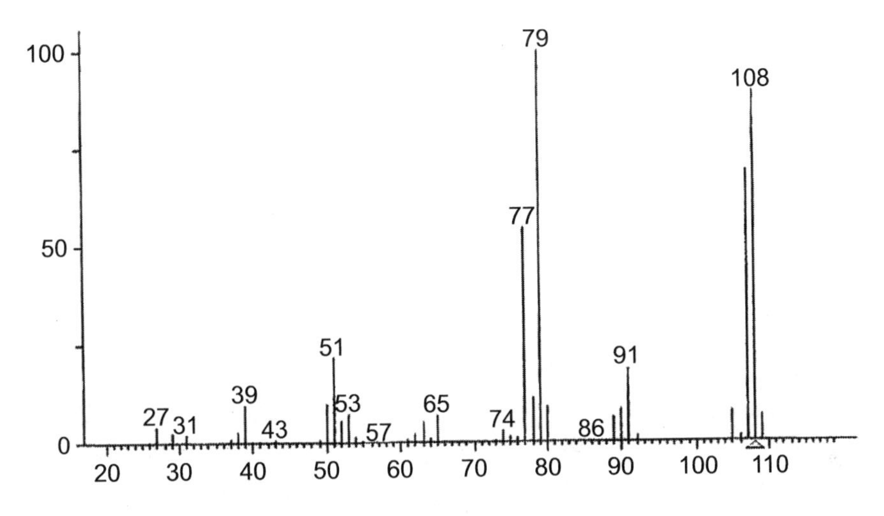

Figure 3.52

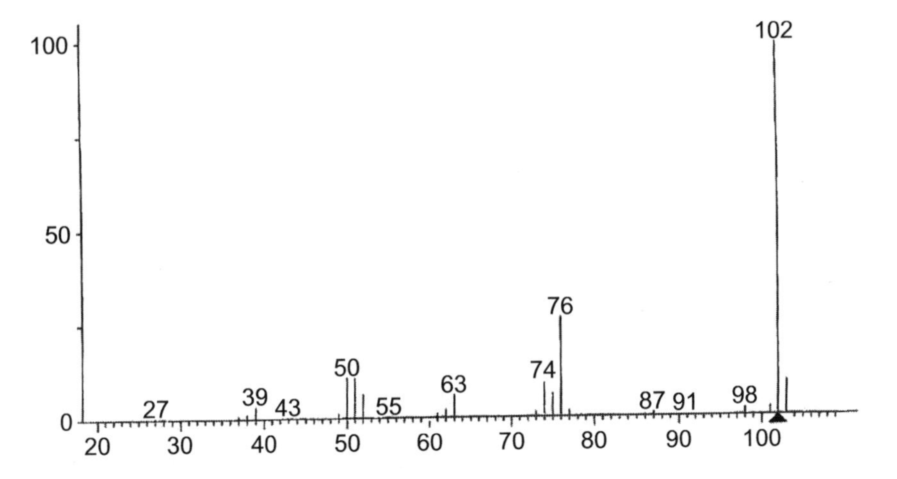

Figure 3.53

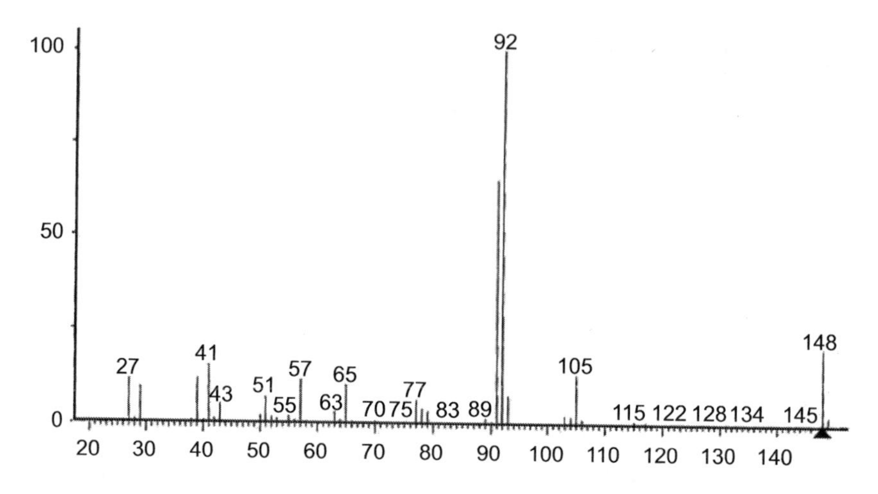

Figure 3.54

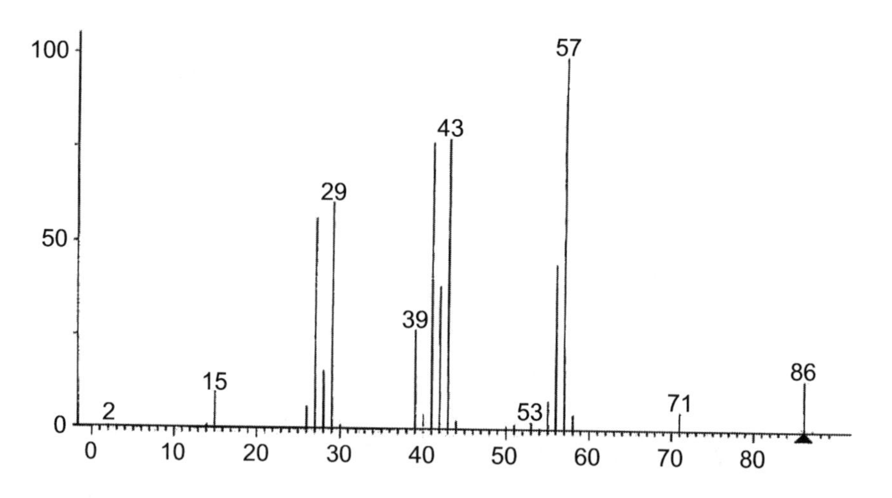

Figure 3.55

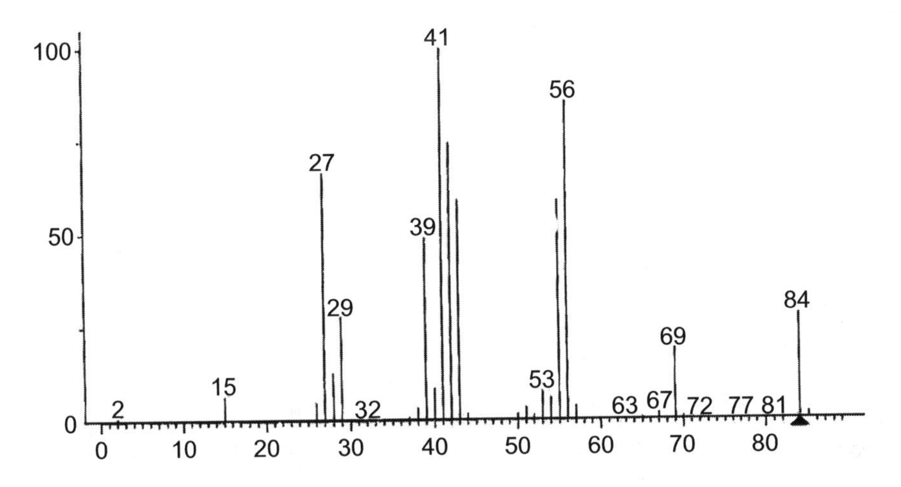

Figure 3.56

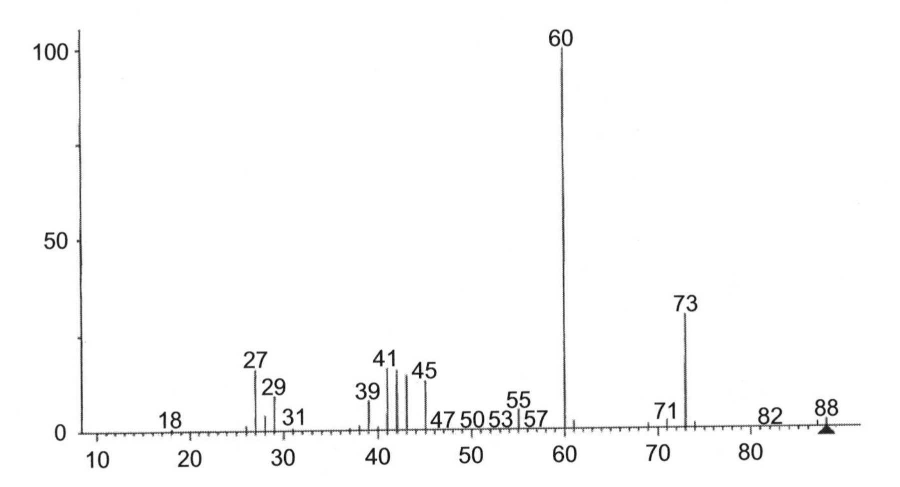

Figure 3.57

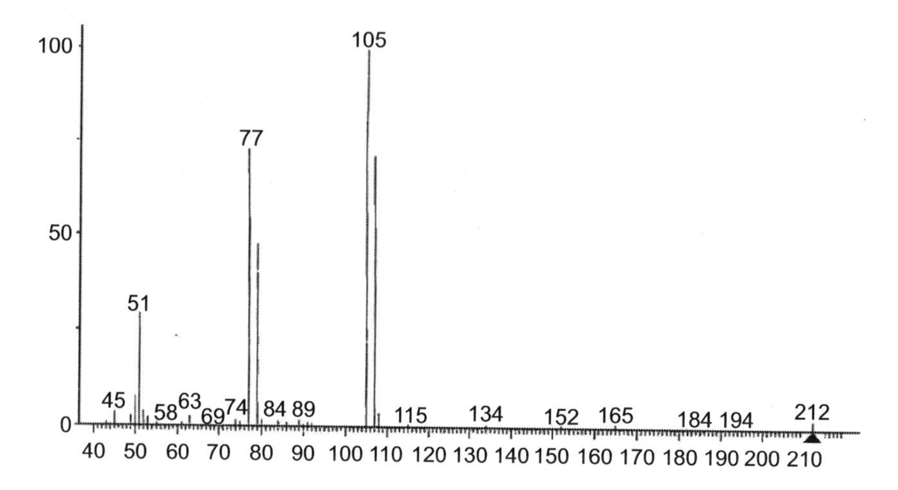

Figure 3.58

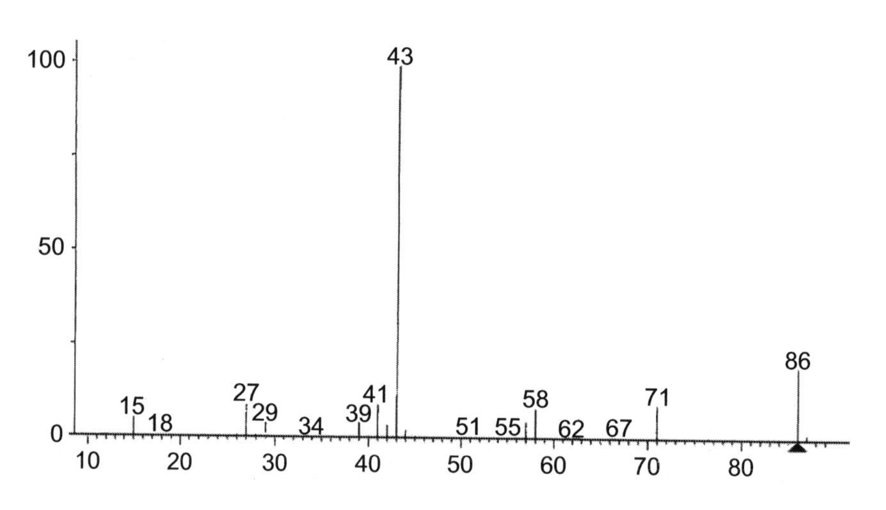

Figure 3.59

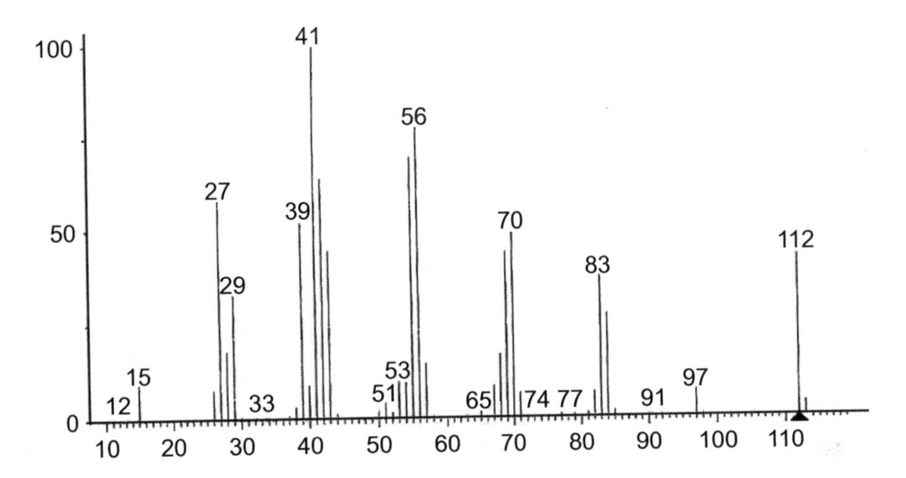

Figure 3.60

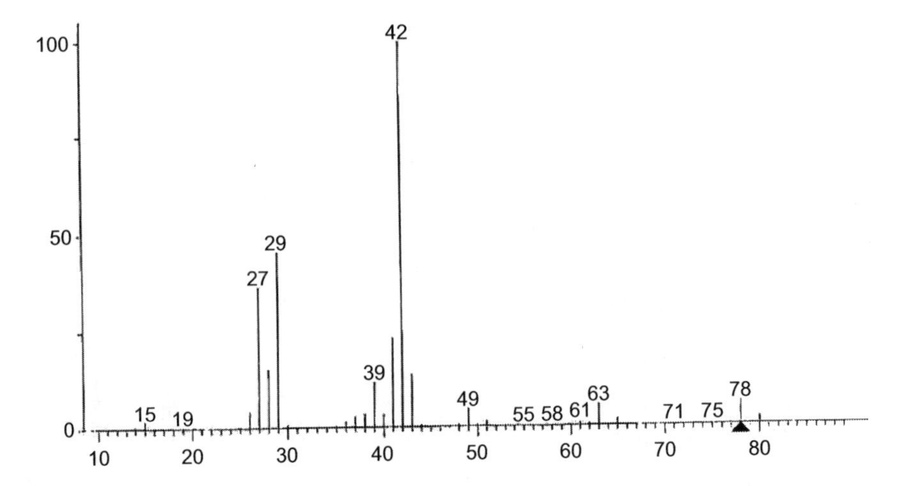

Figure 3.61

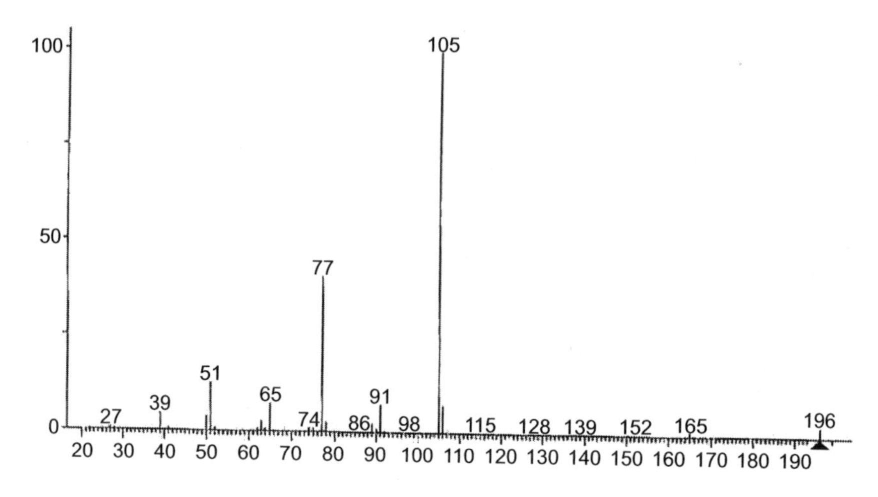

Figure 3.62

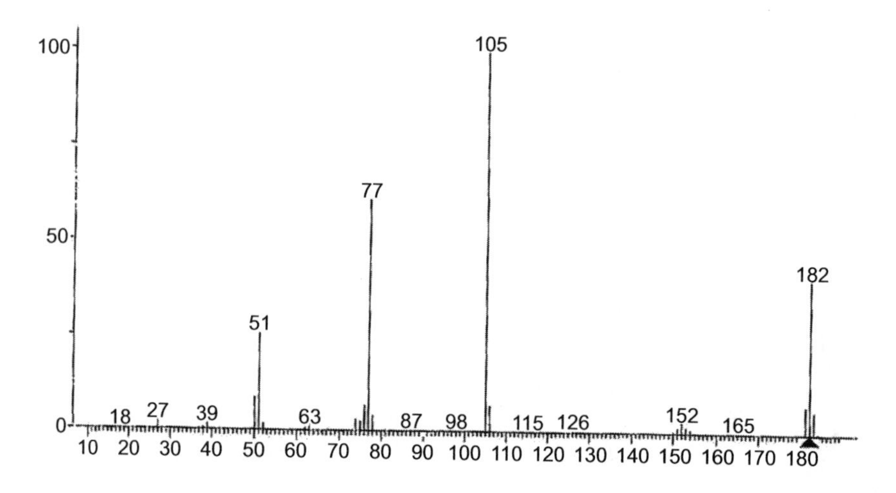

Figure 3.63

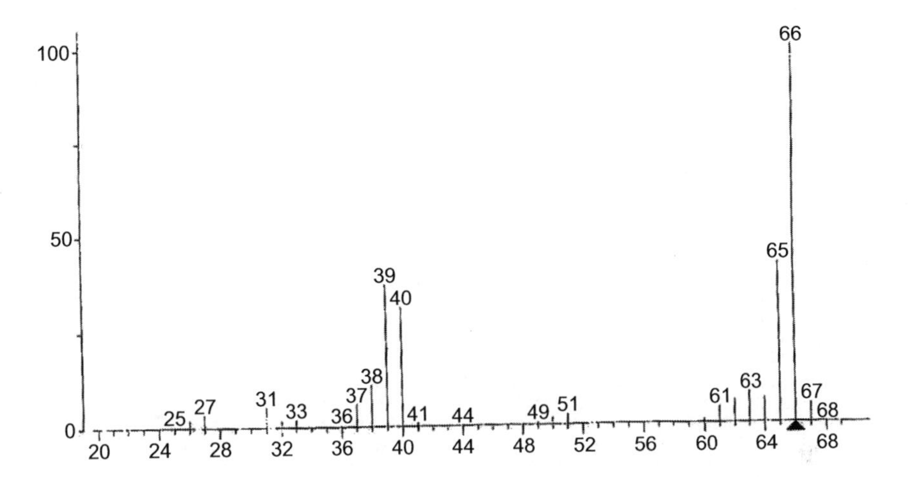

Figure 3.64

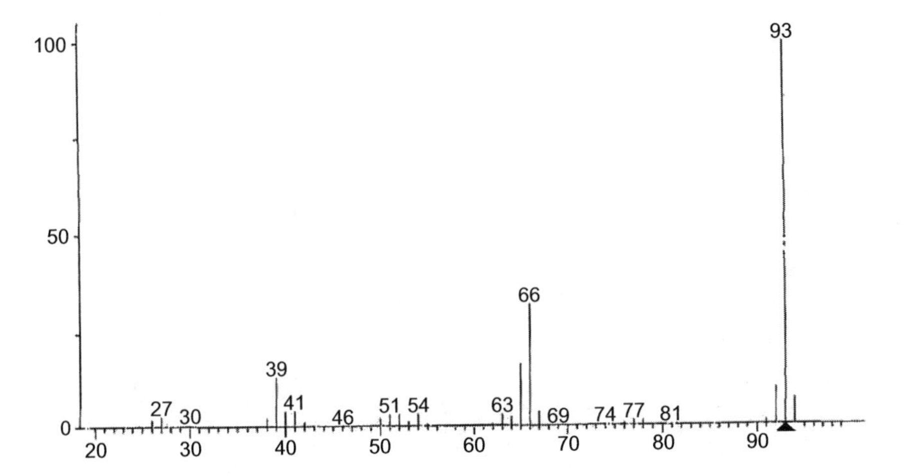

Figure 3.65

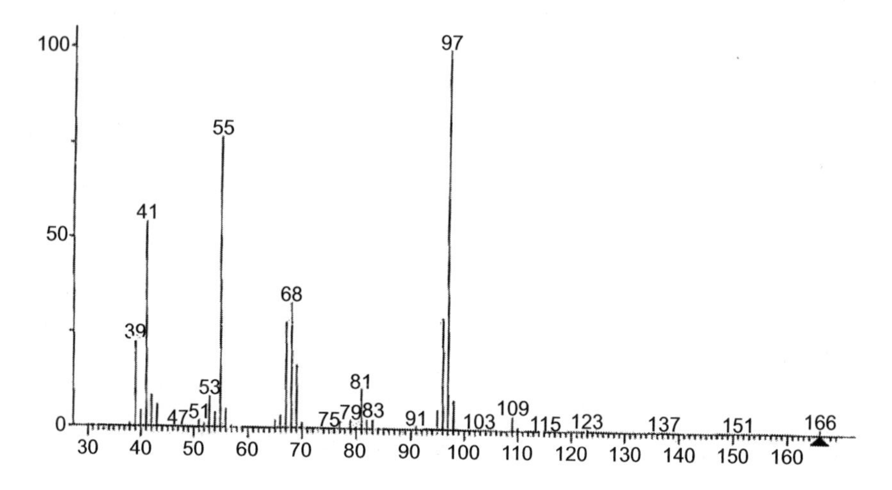

Figure 3.66

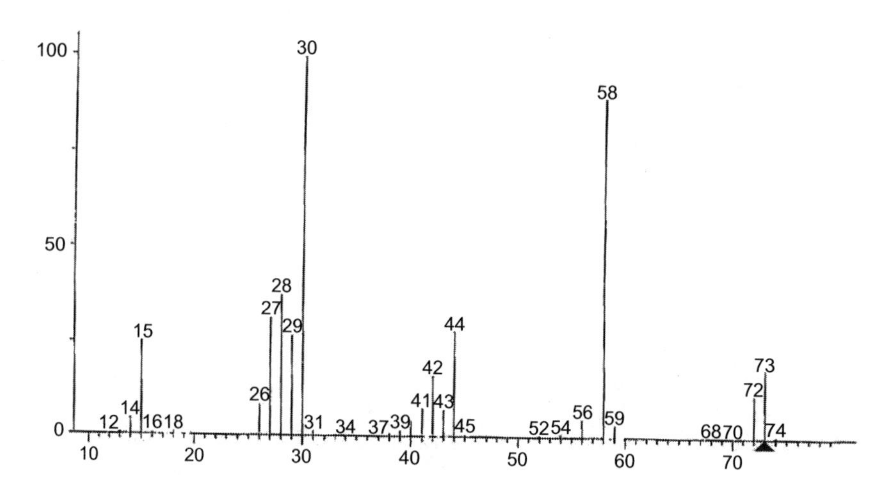

Figure 3.67

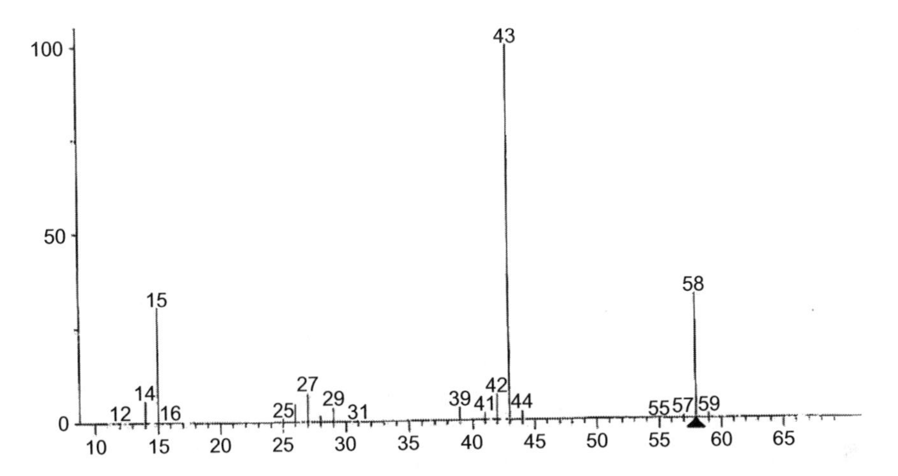

Figure 3.68

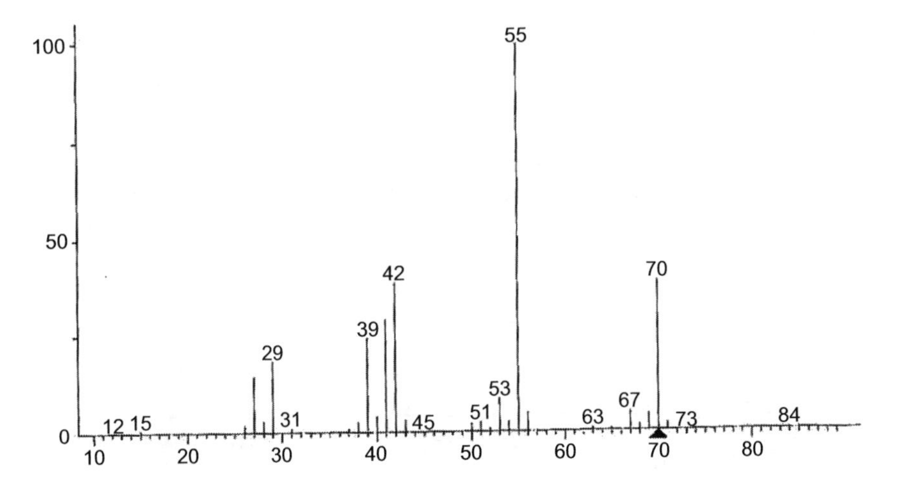

Figure 3.69

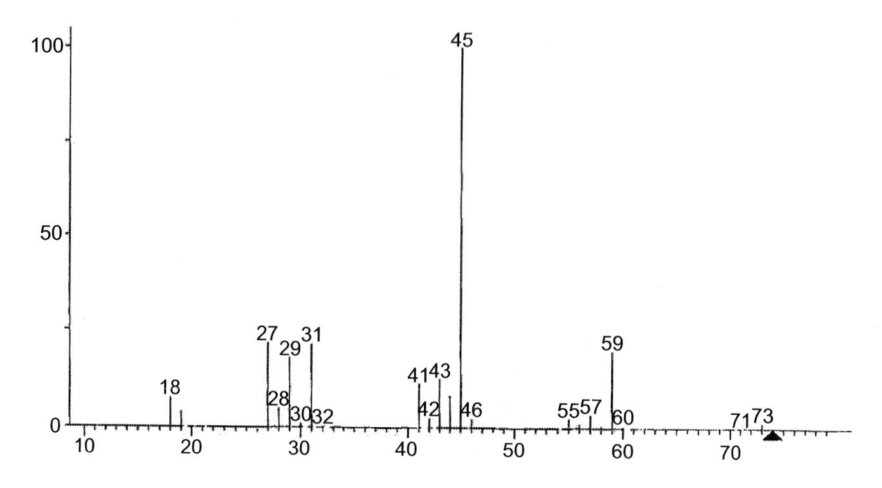

Figure 3.70

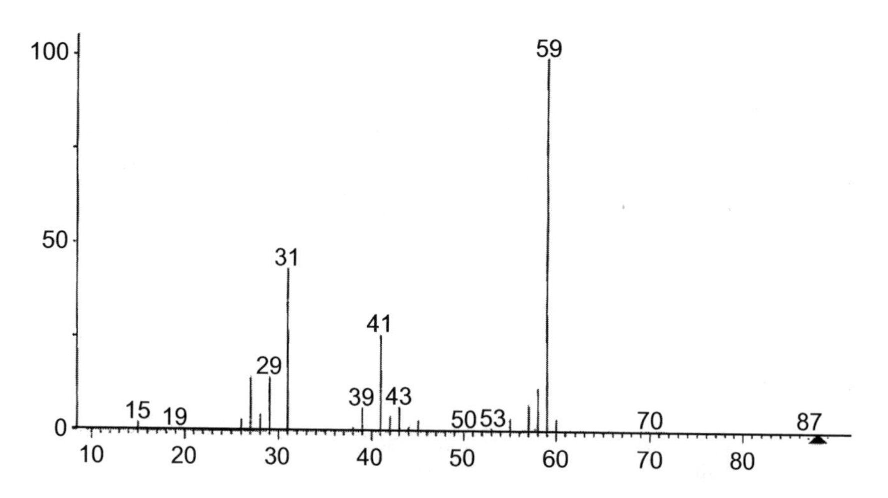

Figure 3.71

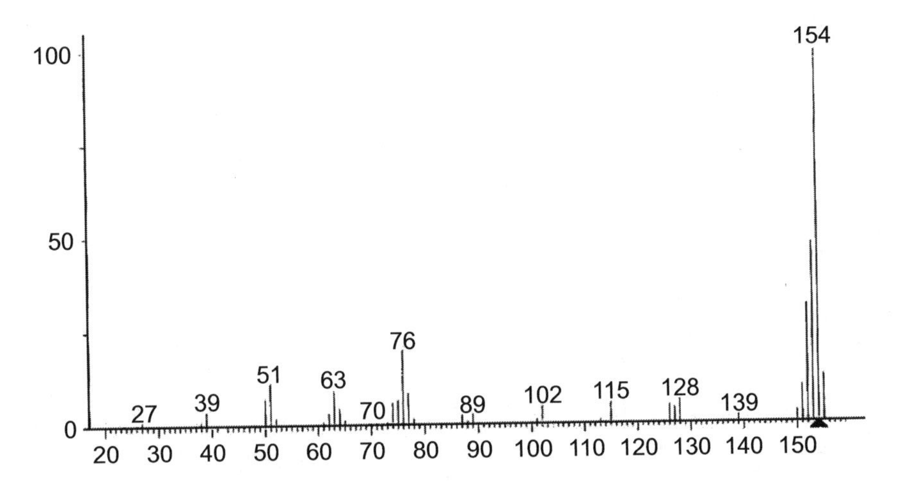

Figure 3.72

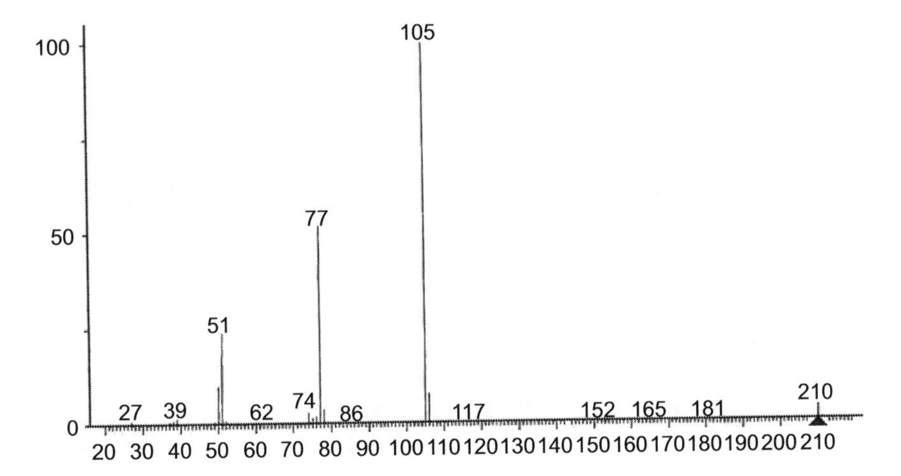

Figure 3.73

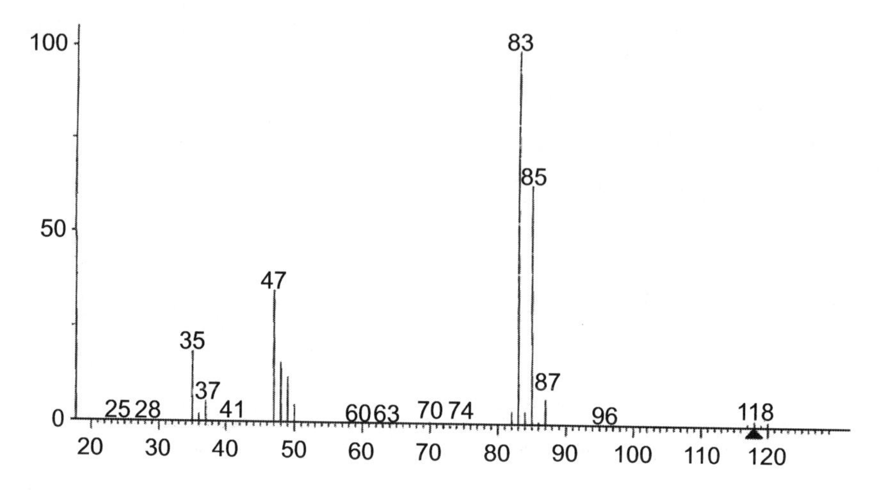

Figure 3.74

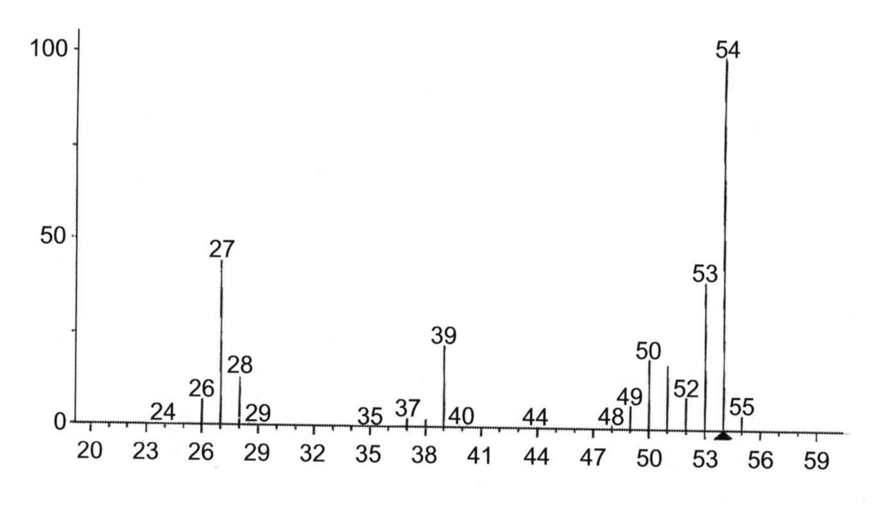

Figure 3.75

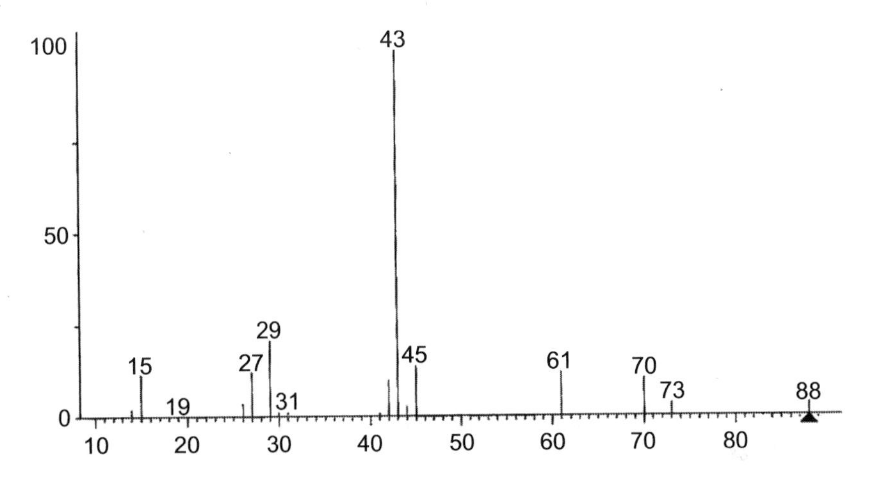

Figure 3.76

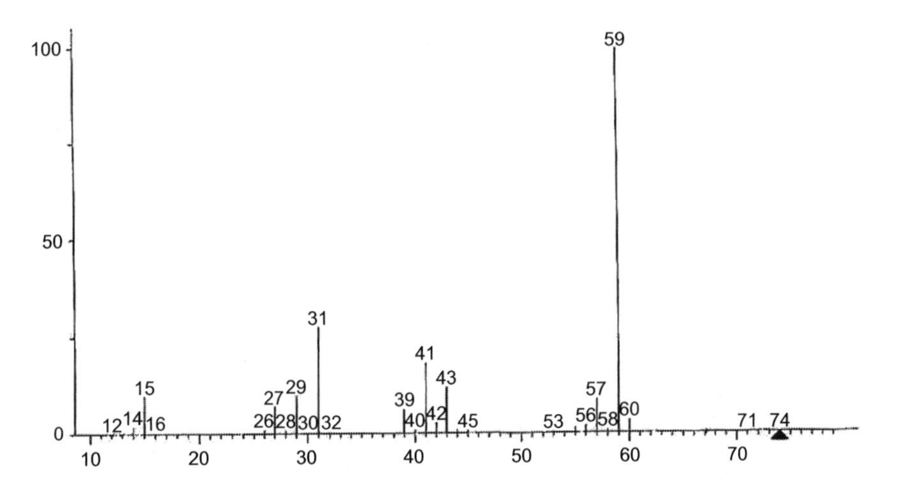

Figure 3.77

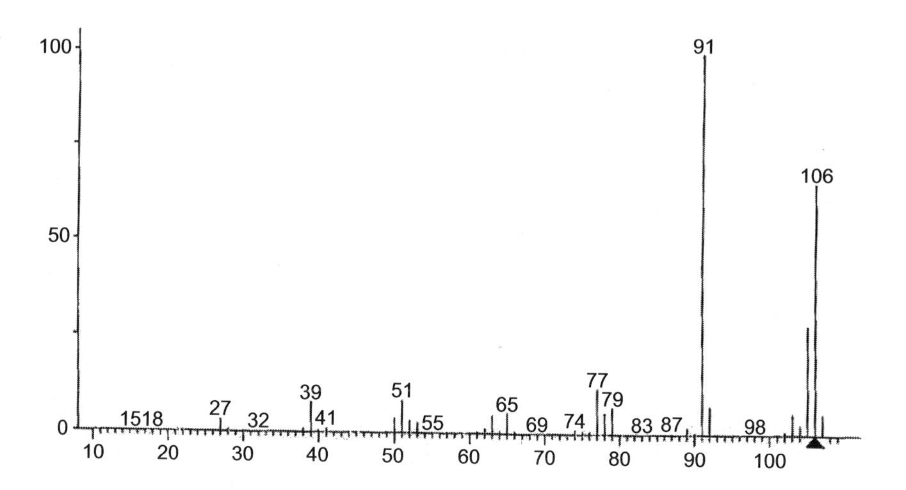

Figure 3.78

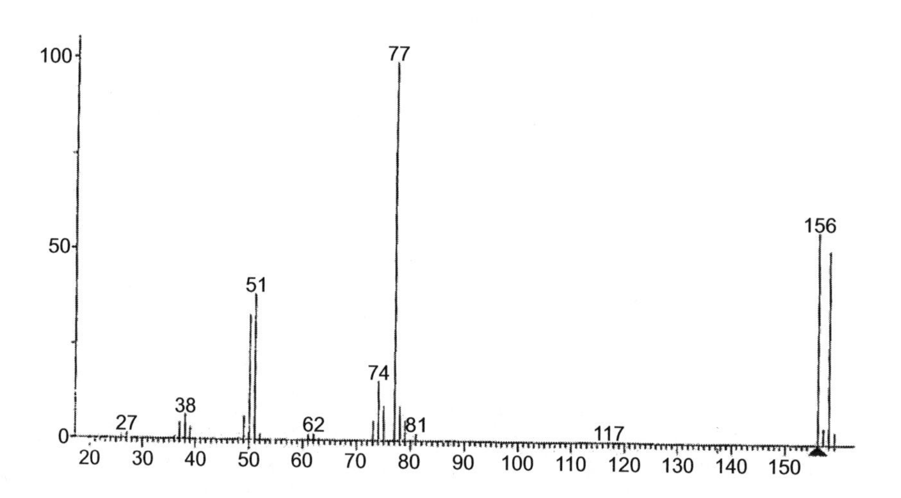

Figure 3.79

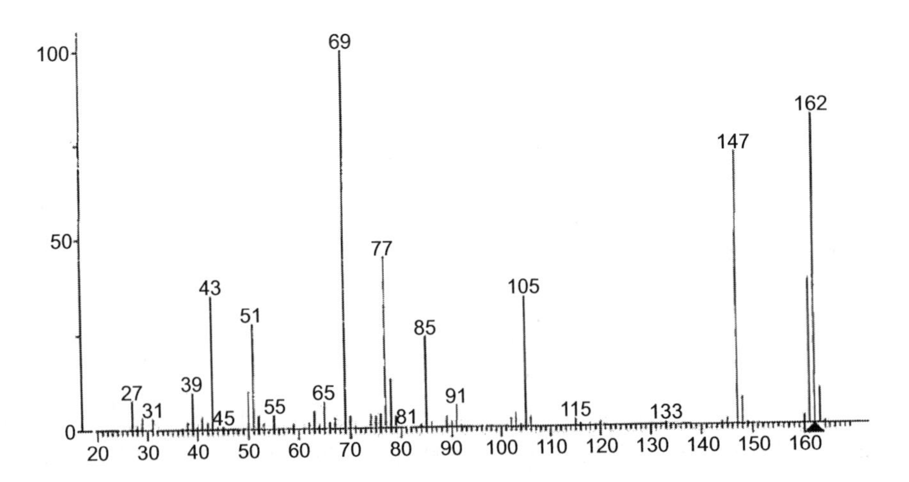

Figure 3.80

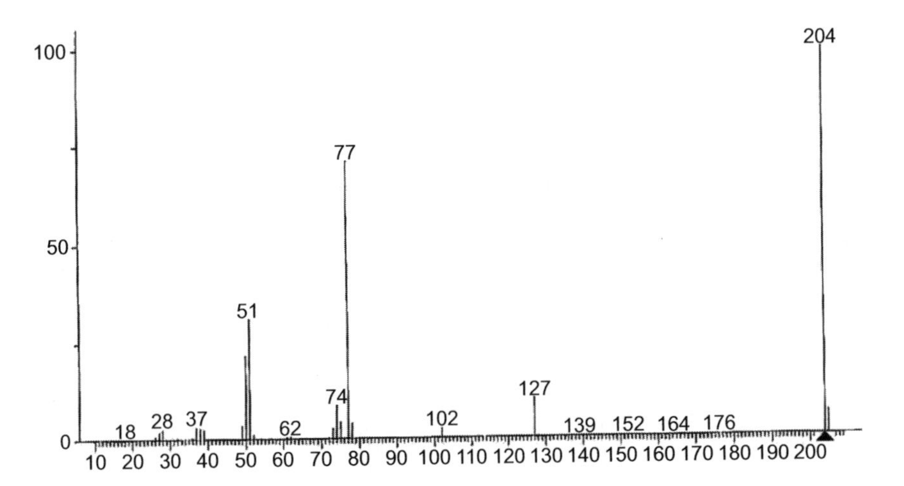

Figure 3.81

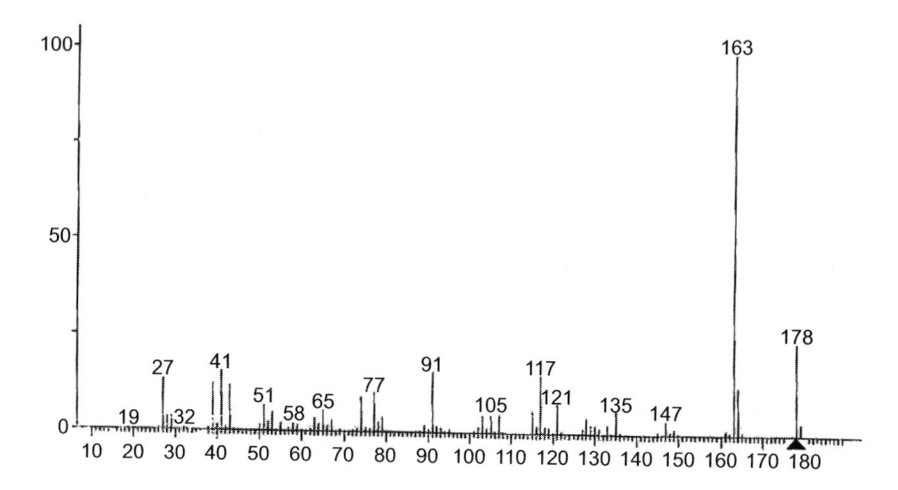

Figure 3.82

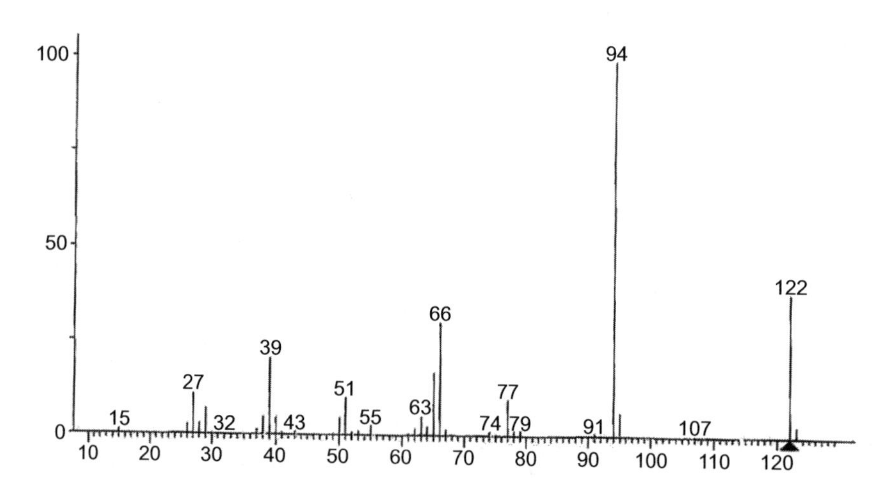

Figure 3.83

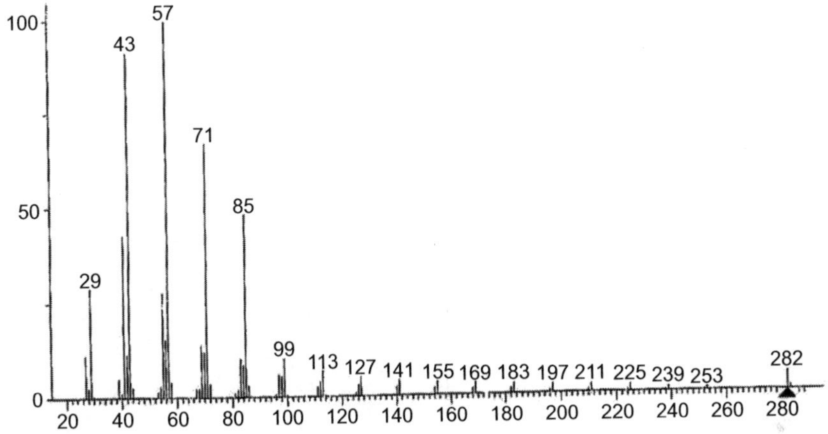

100
50
0
57
43
71
85
29
99
113 127 141 155 169 183 197 211 225 239 253
282
20 40 60 80 100 120 140 160 180 200 220 240 260 280

Figure 3.84

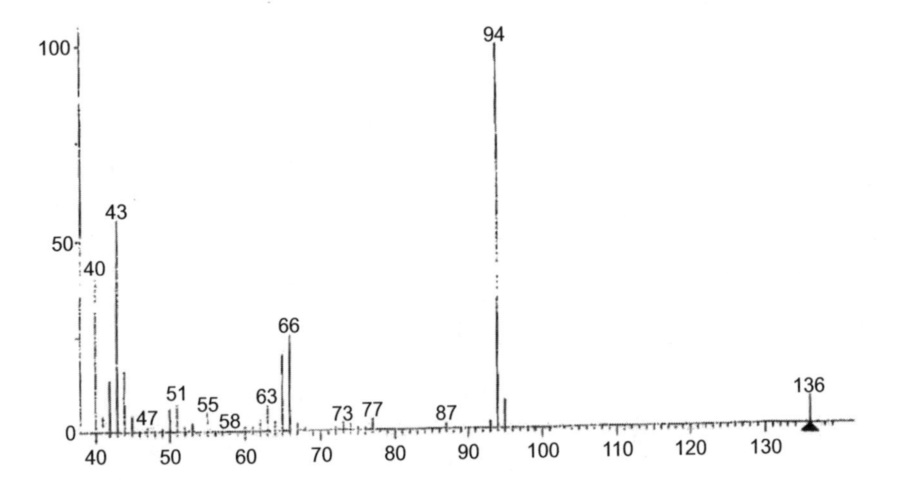

Figure 3.85

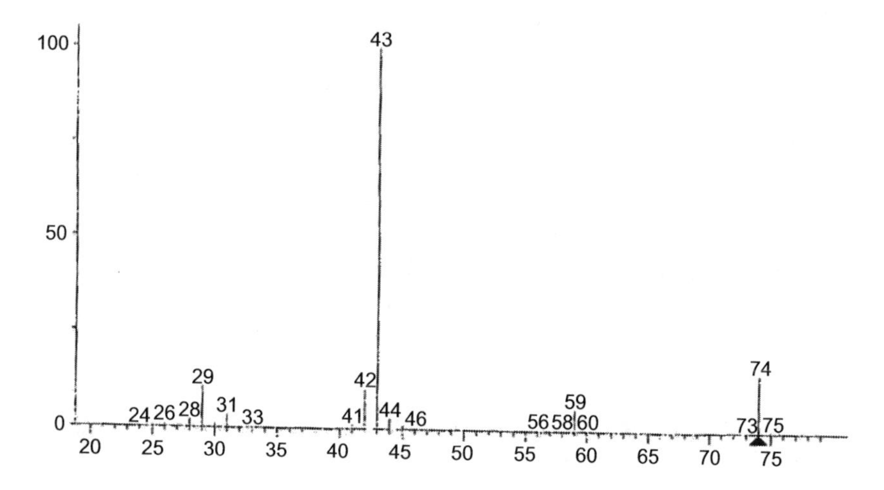

Figure 3.86

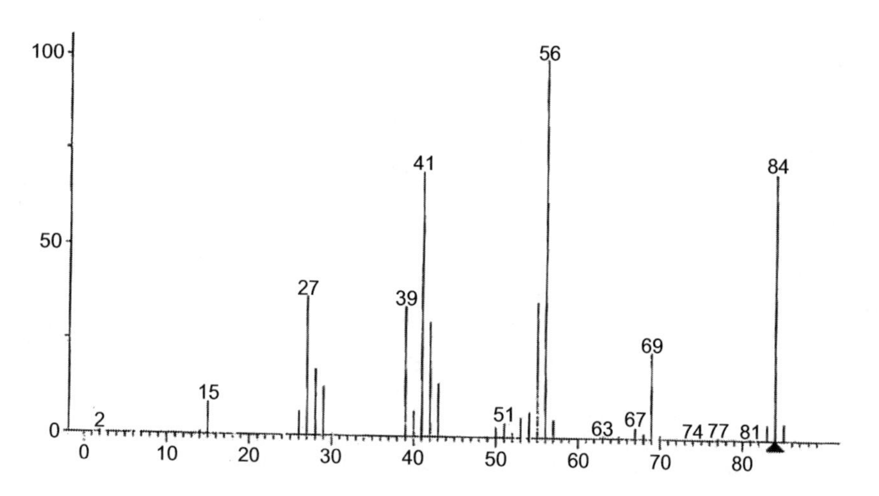

Figure 3.87

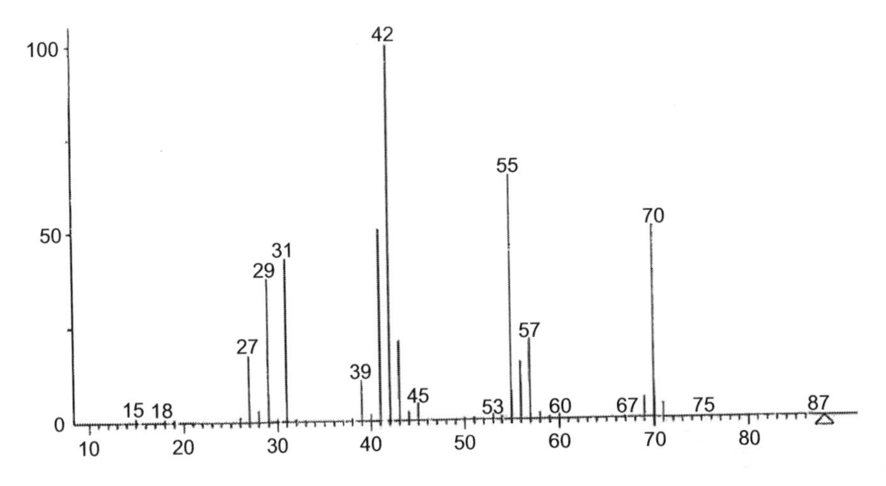

Figure 3.88

3.6 On the Stability of Molecules

It is sometimes possible to obtain a rough indication of the relative stability of molecules by comparing their mass spectra. For instance, consider the mass spectra in Fig. 3.89 (hexanedioic acid) and Fig. 3.90 (nicotine).

In hexanedioic acid, Fig. 3.89, there are quite a few fragments with large and medium abundances, suggesting lots of bond breaking occurs during the fragmentation process. However, in nicotine (Fig. 3.90), there are very few abundant fragments, suggesting much less bond breaking. In fact, in nicotine, the only major fragment is located at $m/z = 84$. It can thus be concluded that nicotine have stronger bonds than hexanedioic acid, and since molecular stability can be related to bond strength, it can be concluded that nicotine is a more stable molecule than hexanedioic acid.

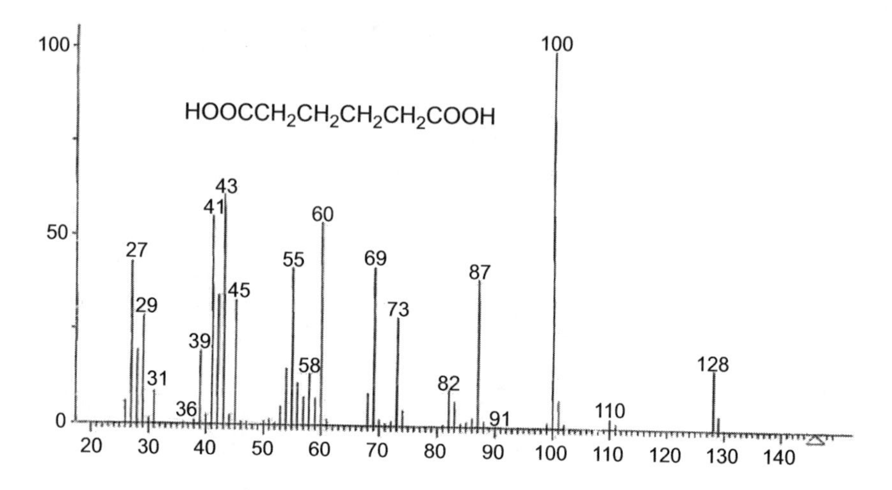

Figure 3.89

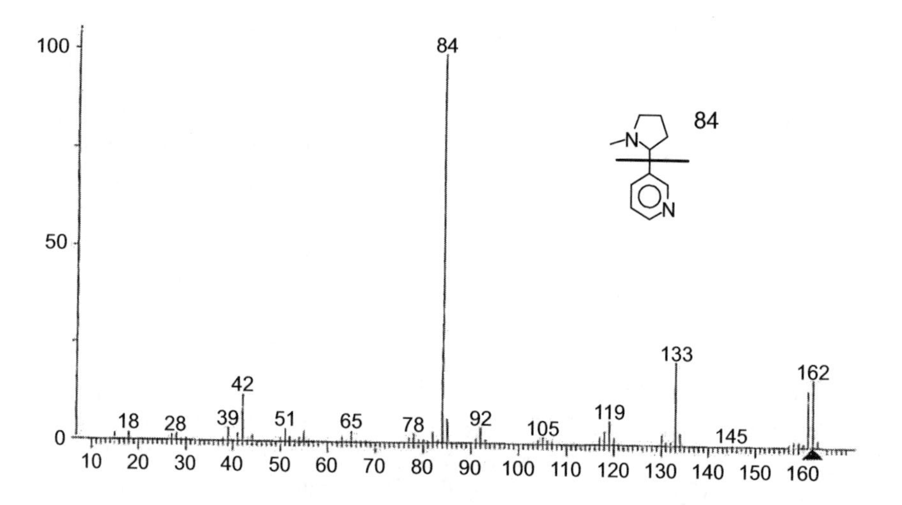

Figure 3.90

3.6.1 Complex Molecules Revisited

(Also see Question 43.)

Consider Figs. 3.91 (cocaine), 3.92 (morphine), 3.93 (methoxsalen), and 3.94 (methenolone enanthate). The four mass spectra represent structures of complex molecules. Let us assume that you did not know the structures and names of the four compounds, then how would you identify them?

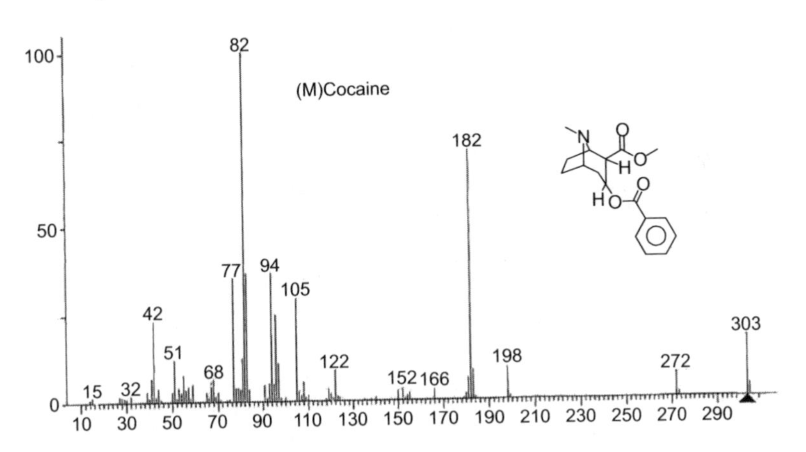

Figure 3.91

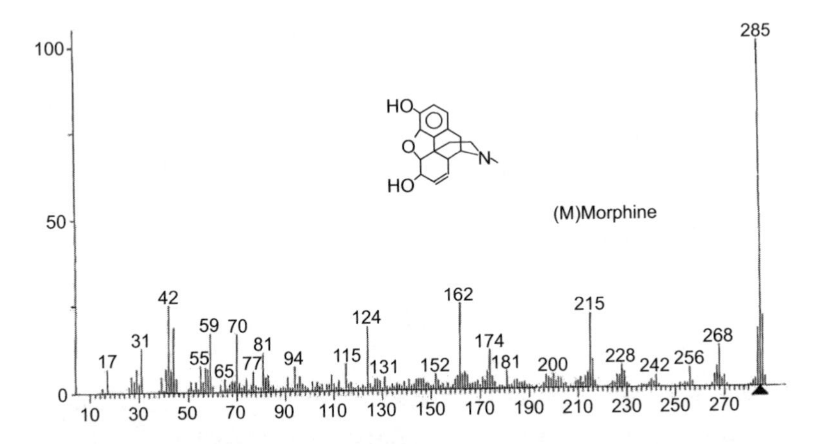

Figure 3.92

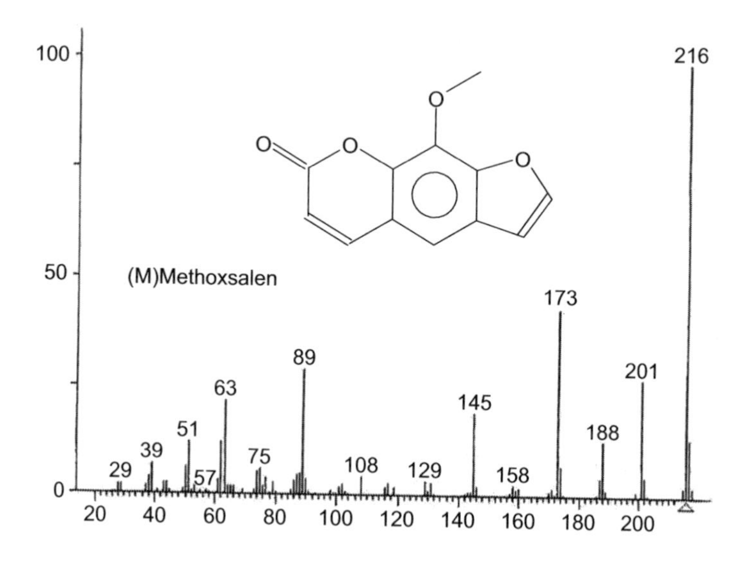

Figure 3.93

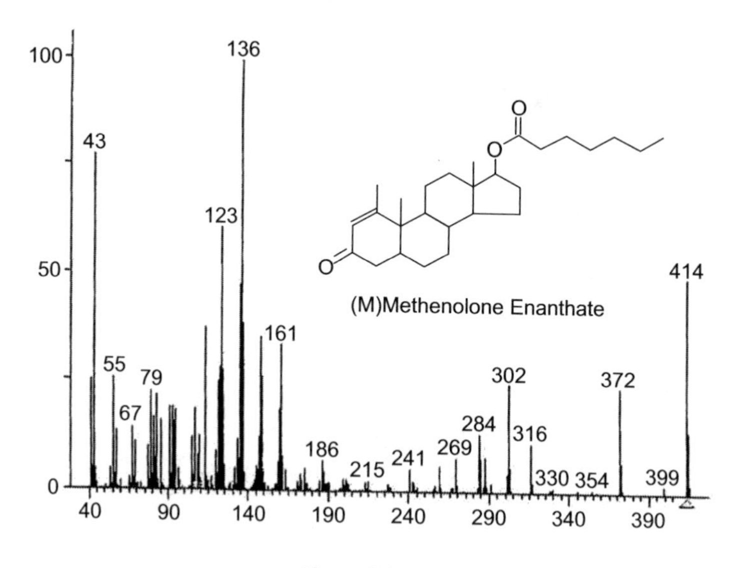

Figure 3.94

The ideal situation is to search the mass spectra of the four structures against a mass spectra Database such as Wiley or NIST. For this to occur, the mass spectra must be in the computer under the proper program, and you must have the database as well as all the necessary supporting software. You have none of this. All you have is a printed copy of the mass spectra of the four compounds and a copy of NIST. What can you do? How can you identify the structures of the four "unknowns"?

Assuming that you know the molecular weights of the four "unknowns," you can carry out a molecular weight search using NIST. This gives 285 hits for the molecular weight of morphine, 303 hits for cocaine, 400 hits for methoxsalen, and 245 hits for methenolone enanthate. Thus, the molecular weight search is not helpful. Using the NIST "search by any peak" method, you are able to obtain some extraordinary results. In all four examples, taking only the two most abundant fragments, the following hits were obtained and they all contained the name of the correct compound.

See the following pages for a more detail explanation.

- Cocaine (3 hits)
- Morphine (1 hit)
- Methosalen (5 hits)
- Methenolone nanthate (3 hits)

Cocaine (3 hits)

Mass	Abundances	Computer hits	Peaks	Hits
82	90–100	701	1	778
182	70–80	83	2	3

Thus, when the NIST Library of Mass Spectra is search against just two mass and their abundances, we get the following hits:
1. Pseudococaine
2. Cocaine
3. Cinnamoylcocaine

Morphine (1 hit)

Mass	Abundances	Computer hits	Peaks	Hits
285	90–100	183	1	718
162	20–25	537	2	1

1. Morphine (only one hit)

Methoxsalen (5 hits)

Mass	Abundances	Computer hits	Peaks	Hits
216	90–100	429	1	429
173	50–60	180	2	5

1. Amino-3-chloro-5-mecaptoacetanilide,
2. 2H-Furo[2,3,h]-1-benzopyran-2-one, 5-methoxy
3. 7H-Furo[3,2,g]-1-benzopyran-7-one, 4-methoxy
4. 2-Carbethoxy-3-hydroxymethyl-5-methoxy-6-methylindole
5. Methoxsalen

Methenolone Enanthate (3 hits)

Mass	Abundances	Computer hits	Peaks	Hits
136	90–100	673	1	937
123	60–700	270	2	3

1. Androsta-5,16-dienol[16,17-b]indolizin-3-beta-ol
2. Methenolone enanthate
3. 4,8,13-Cyclotetradecatriene-1,3-diol,1,5,9-trimethyl-12-(1-methylethyl]-

(63) Using the NIST "search by any peak" program, make correct structural assignments for the six mass spectra shown in Figs. 3.95, 3.96, 3.97, 3.98, 3.99, and 3.100.

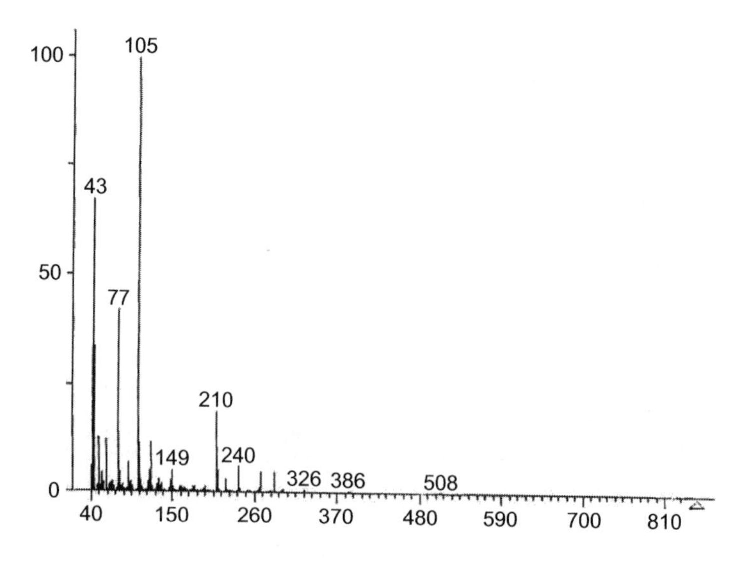

Figure 3.95

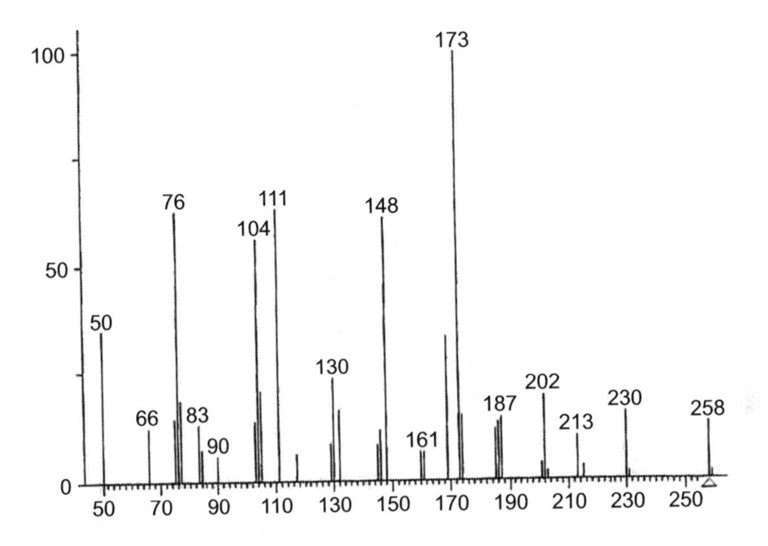

Figure 3.96

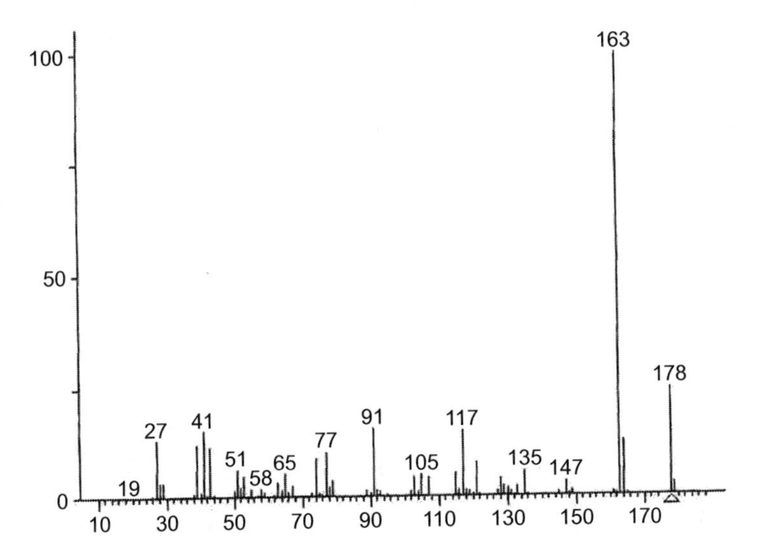

Figure 3.97

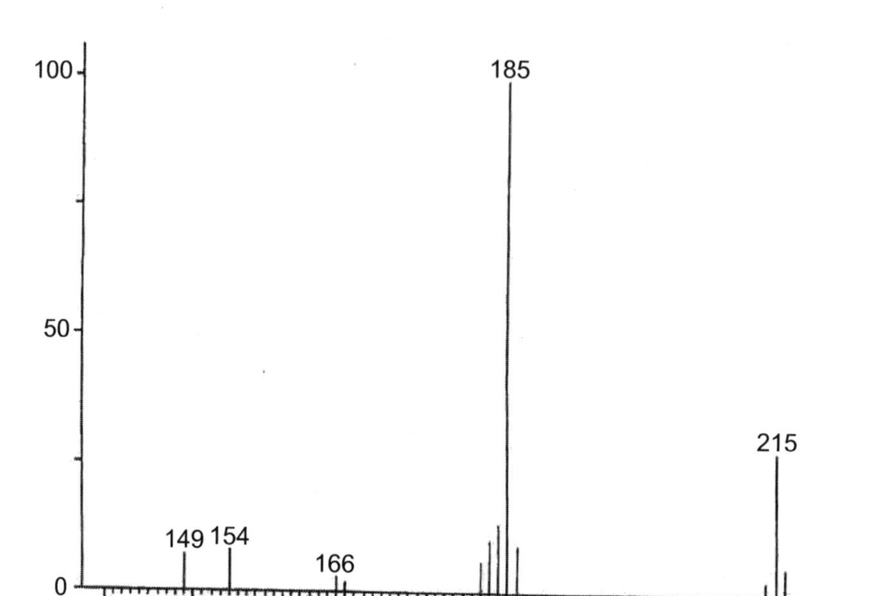

Figure 3.98

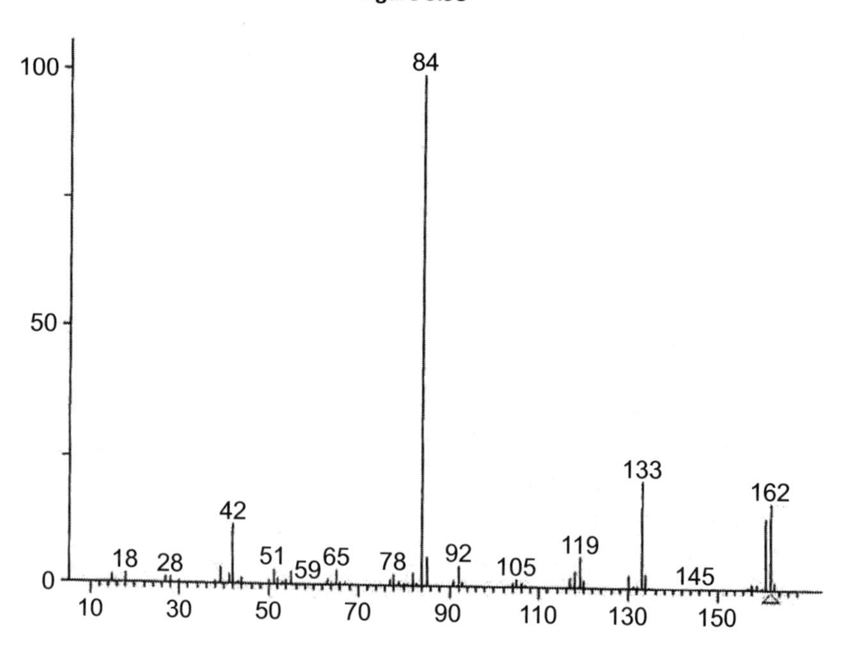

Figure 3.99

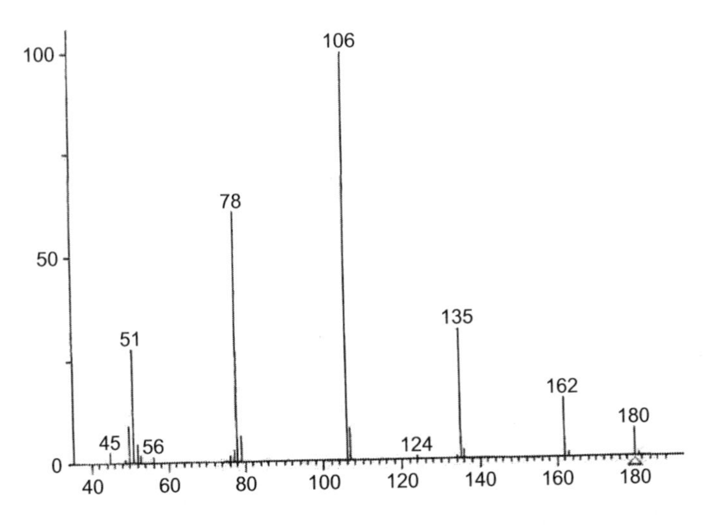

Figure 3.100

3.7 On Detecting Iodine in the Molecule

Iodine is a monoisotopic element, thus its determination in an unknown molecule is different from that of chlorine or bromine. In some compounds containing iodine (not all), the atomic weight of iodine is observed at $m/z = 127$. Some examples are shown in Figs. 3.101, 3.102, 3.103, and 3.104.

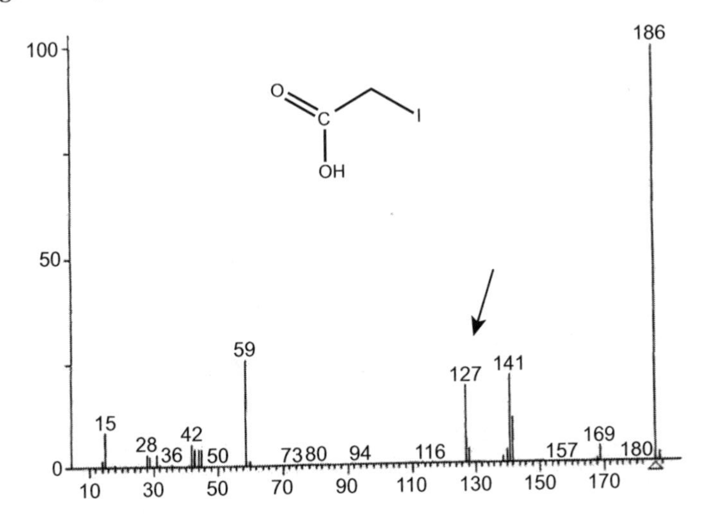

Figure 3.101

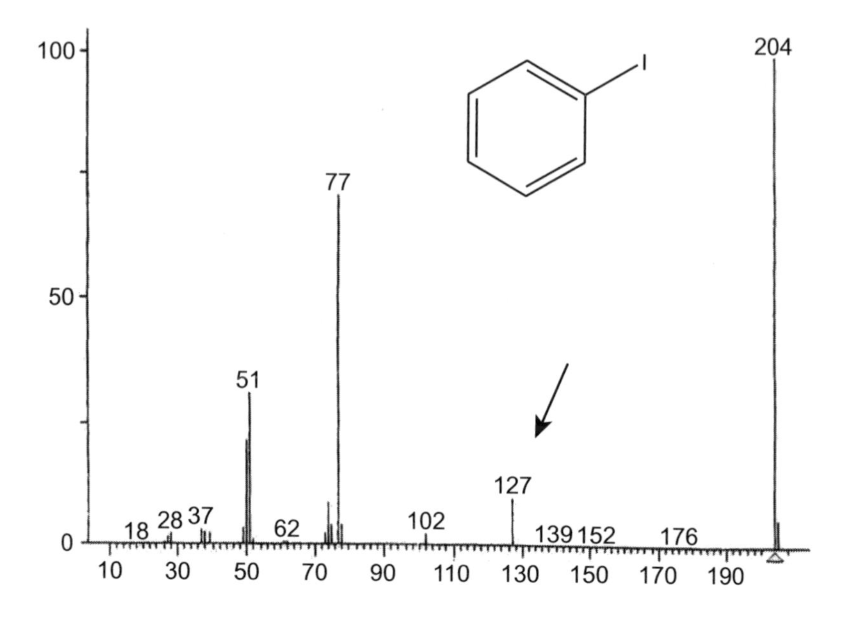

Figure 3.102

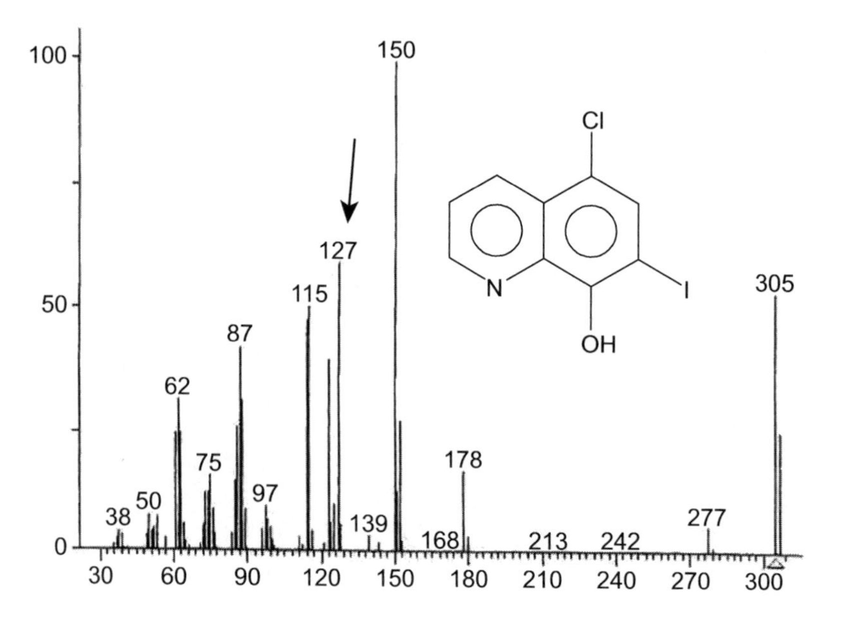

Figure 3.103

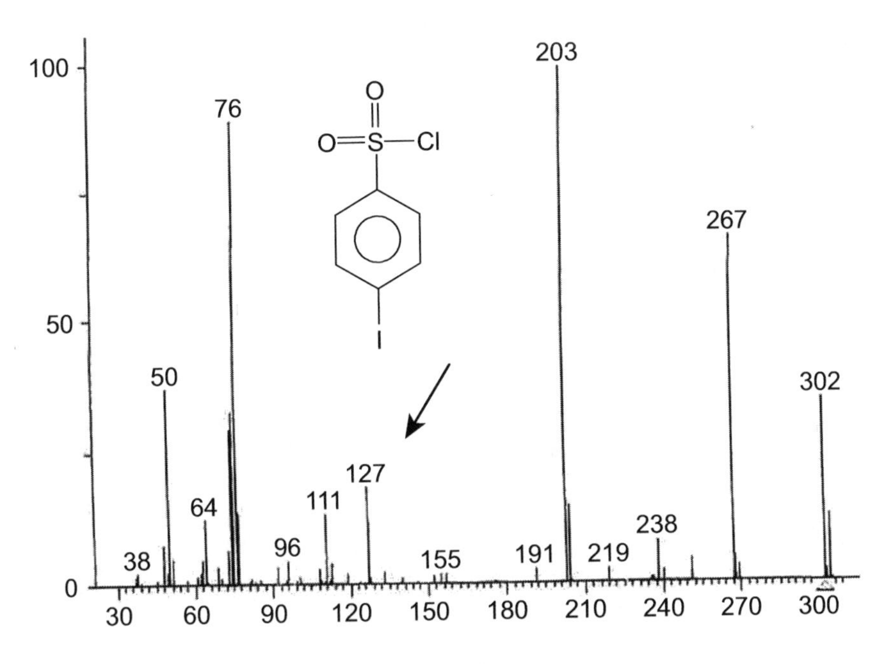

Figure 3.104

3.8 More on Non-Branched Hydrocarbons

The mass spectra of non-branched hydrocarbons are characterized by a series of fragments increasing by a mass of 14 (CH_2), called the Hydrocarbon Cluster. Examples are shown in Figs. 3.105 and 3.106. In Fig. 3.105, masses are observed at 29, 43, 57, 71, 85, 99, 113, 127, etc. If the molecular weight is known, the number of carbons in the hydrocarbon may be obtained by subtracting a hydrogen atom from each of the end methyl groups, followed by division by 14. This is illustrated below for the structures represented by hexacosane and nonane (Figs. 3.105 and 3.106).

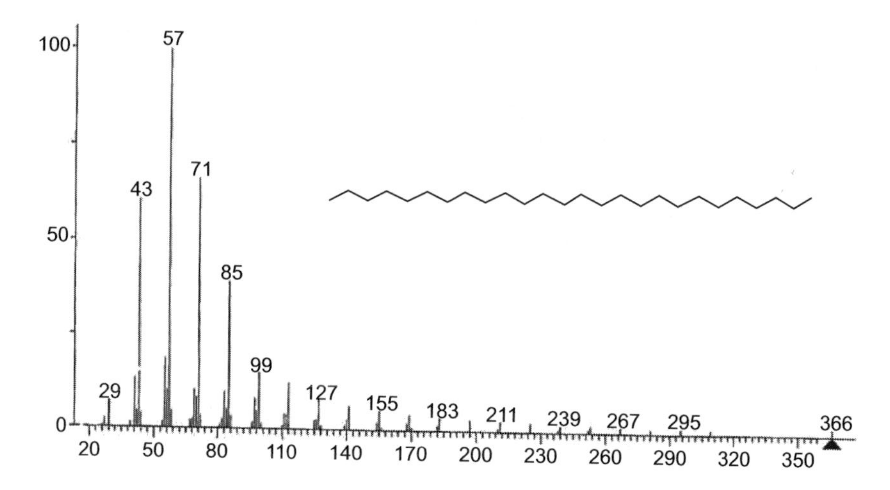

Figure 3.105

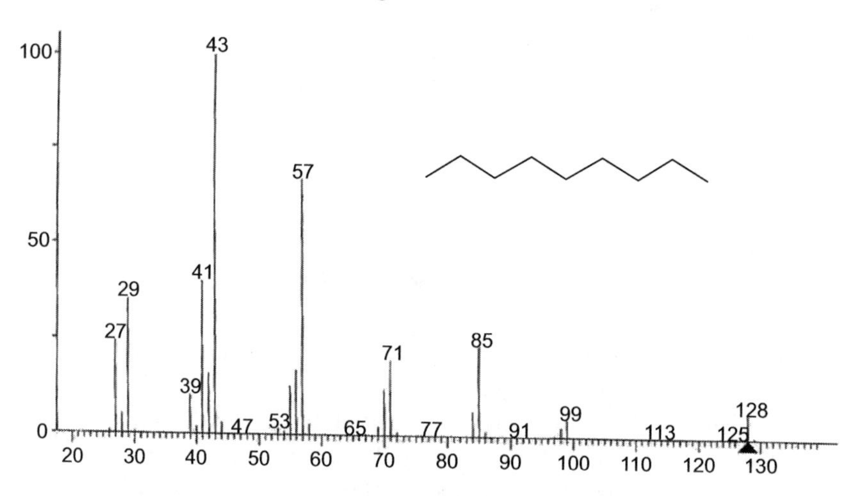

Figure 3.106

3.9 On the Loss of Fragments from the Molecular Ion

Among the most common fragments lost from the molecular ion are: H (1), CH_3 (15), H_2O (18), HF (20), and CO_2 (44). Some examples are shown below.

3.9.1 Loss of H

The loss of H, as we have seen, can be observed in the formation of the tropylium ion ($m/z = 91$) from toluene (Fig. 3.108). Many structures with a methyl phenyl group will lose (H) from M^+ to form a tropylium-like ion as shown below in Fig. 3.107.

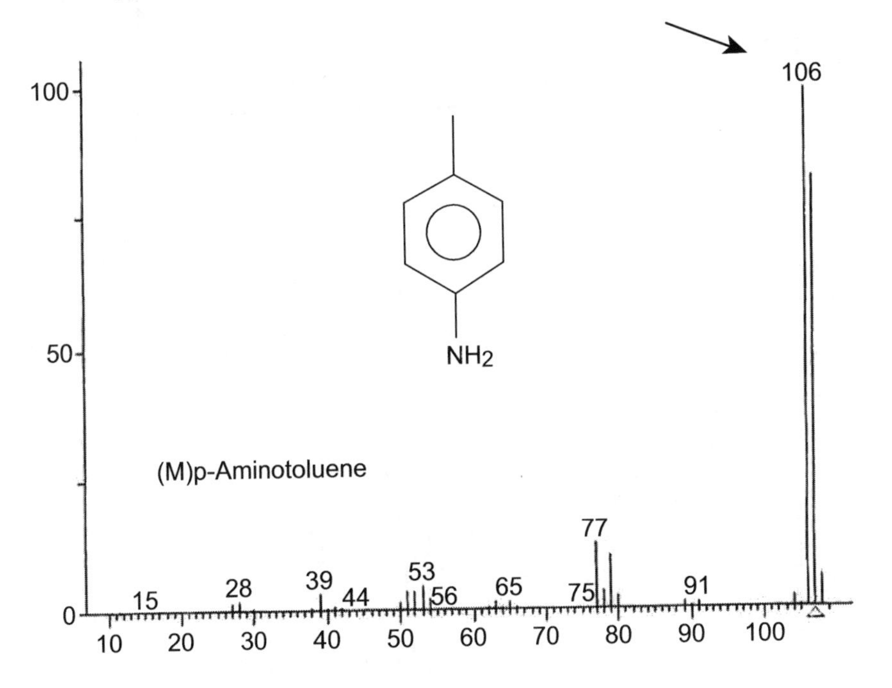

Figure 3.107

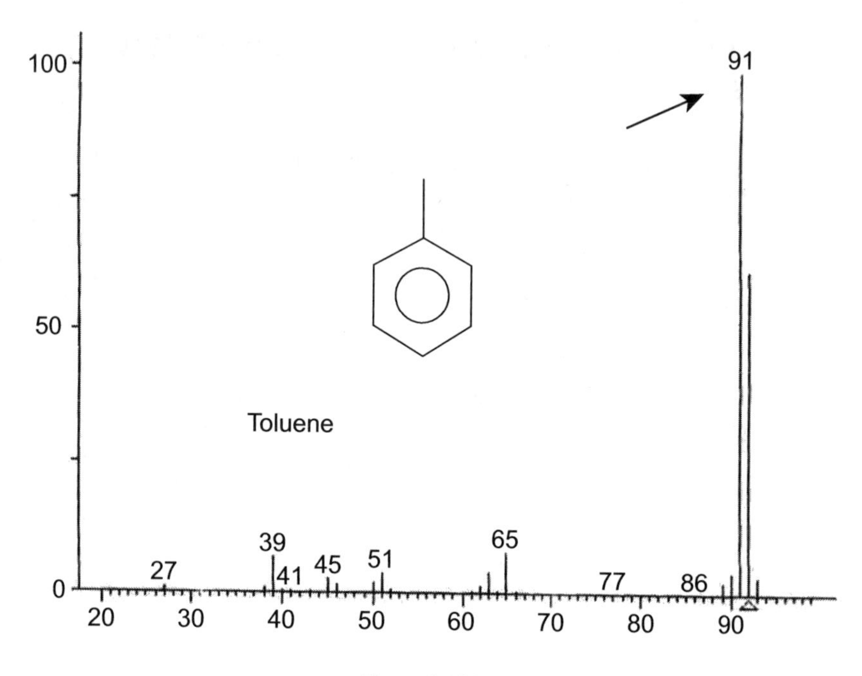

Figure 3.108

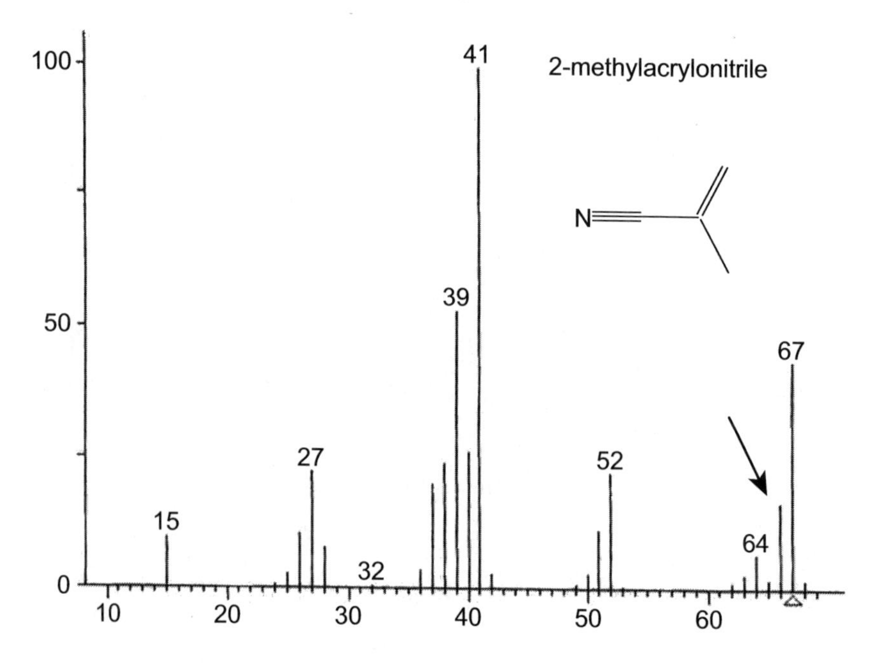

Figure 3.109

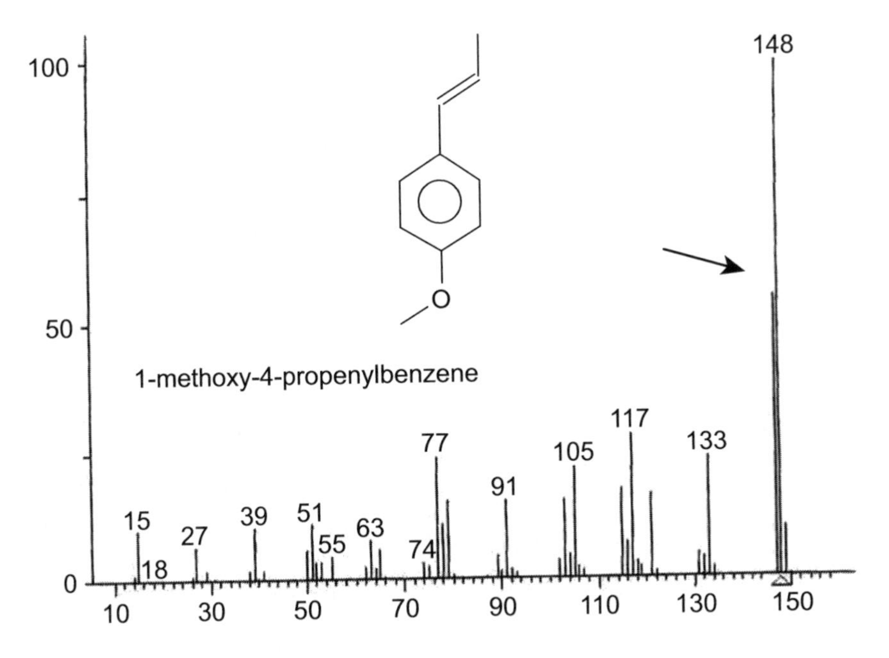

Figure 3.110

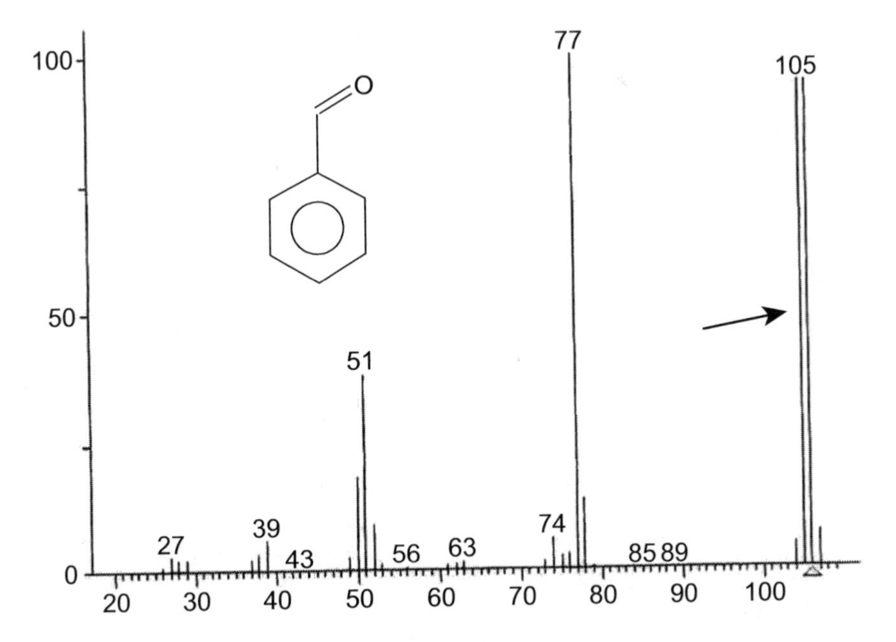

Figure 3.111

3.10 Loss of a methyl group

Examples of the loss of a methyl group (Mass 15) from the molecular ion can be shown in Figs. 3.112, 3.113, 3.114, and 3.115.

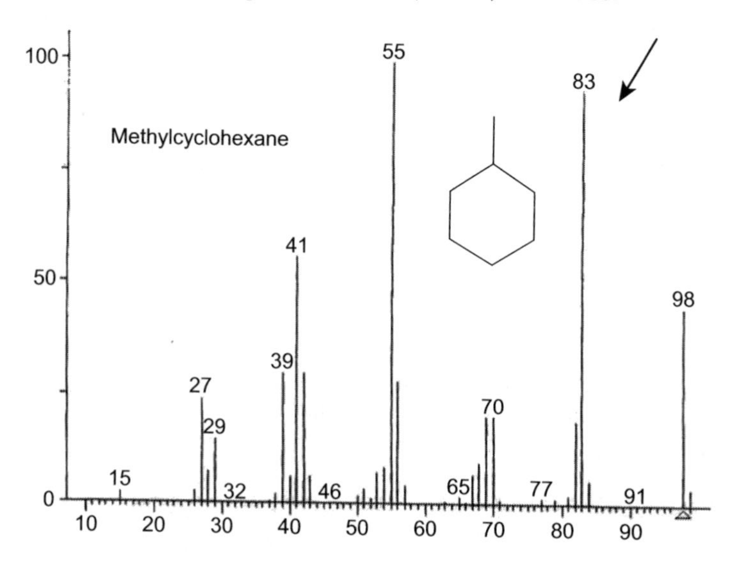

Figure 3.112

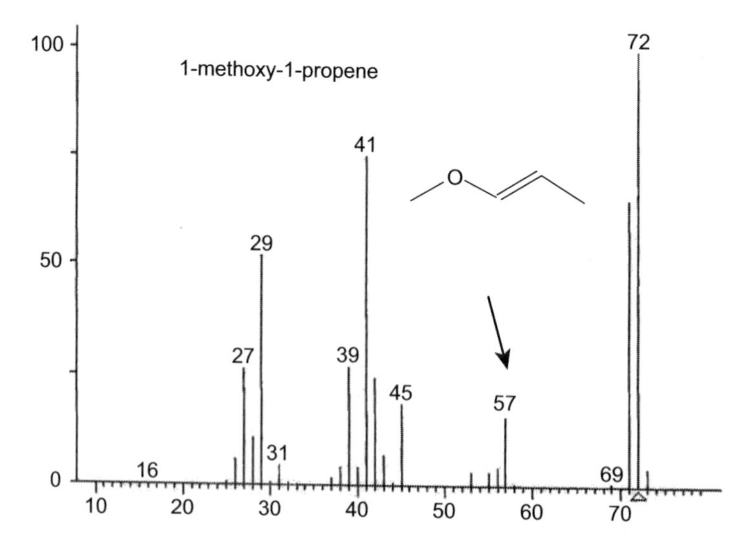

Figure 3.113

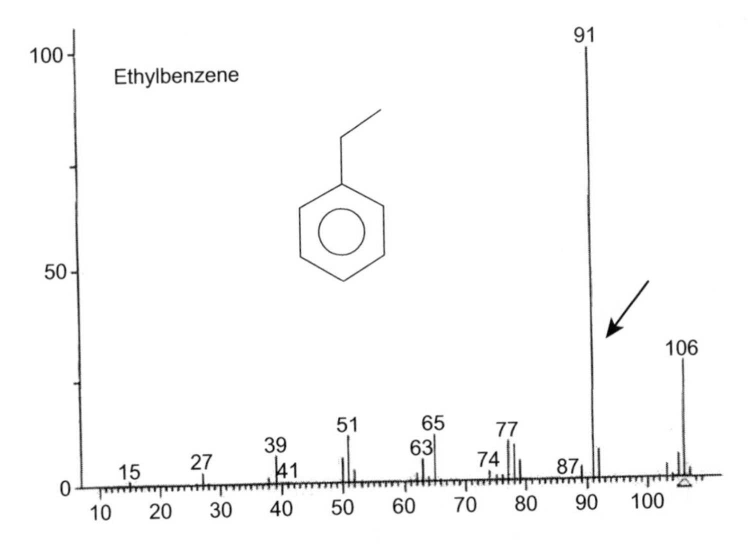

Figure 3.114

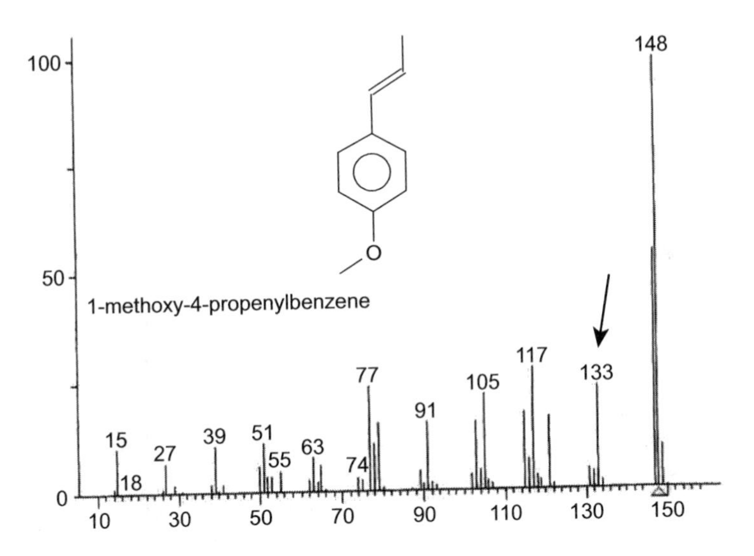

Figure 3.115

3.11 Loss of HF from the Molecular Ion

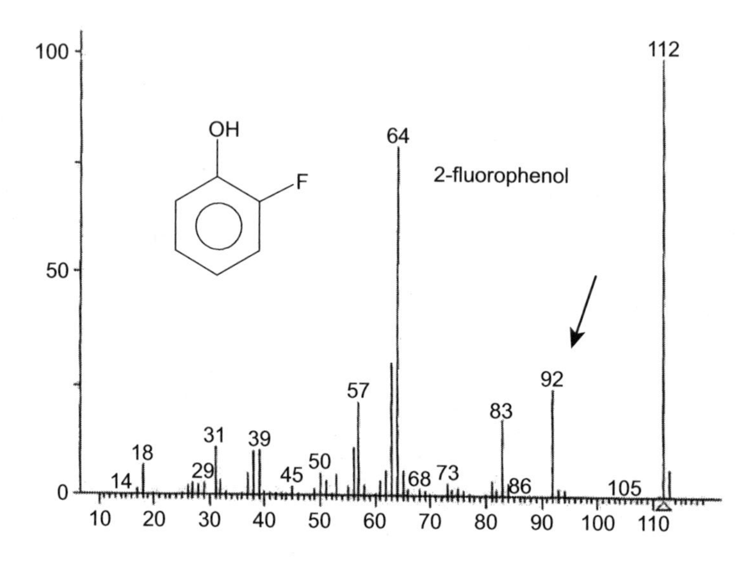

Figure 3.116

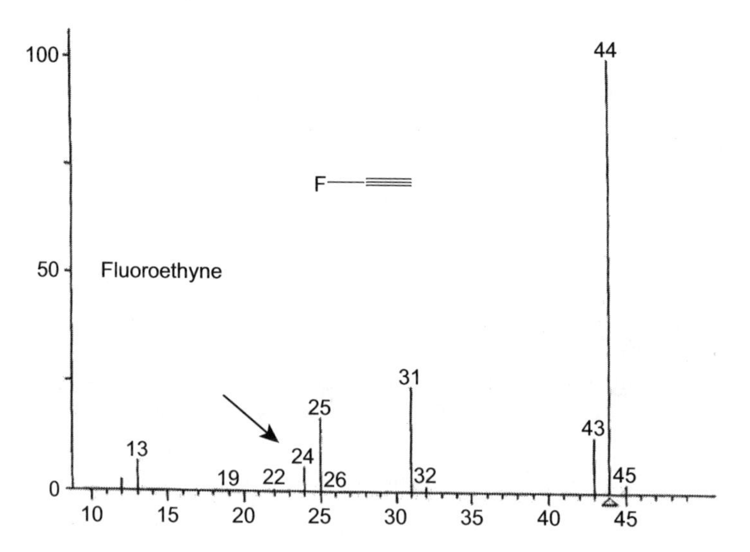

Figure 3.117

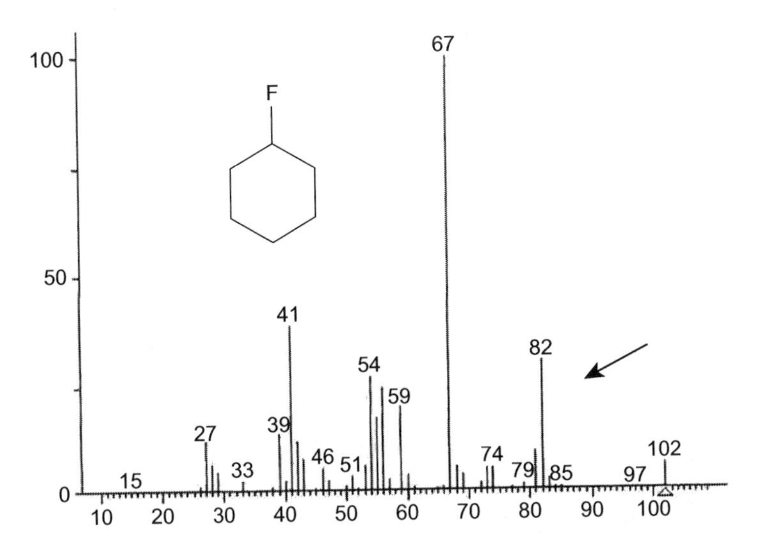

Figure 3.118

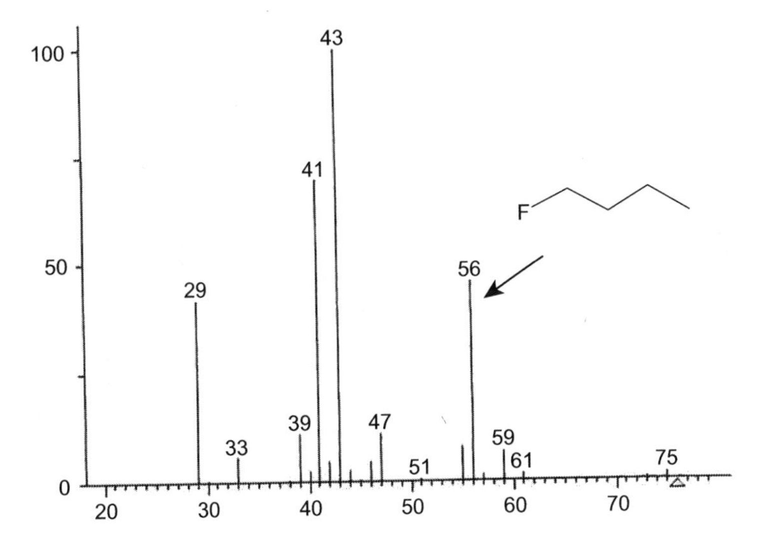

Figure 3.119

3.12 On the Loss of Water from Alcohols

Some examples of the loss of water from alcohols are shown in Figs. 3.120 and 3.121.

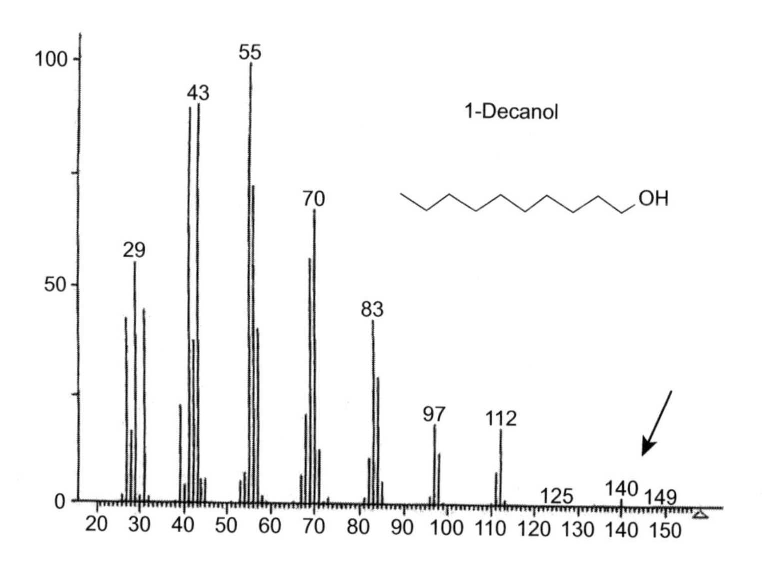

Figure 3.120

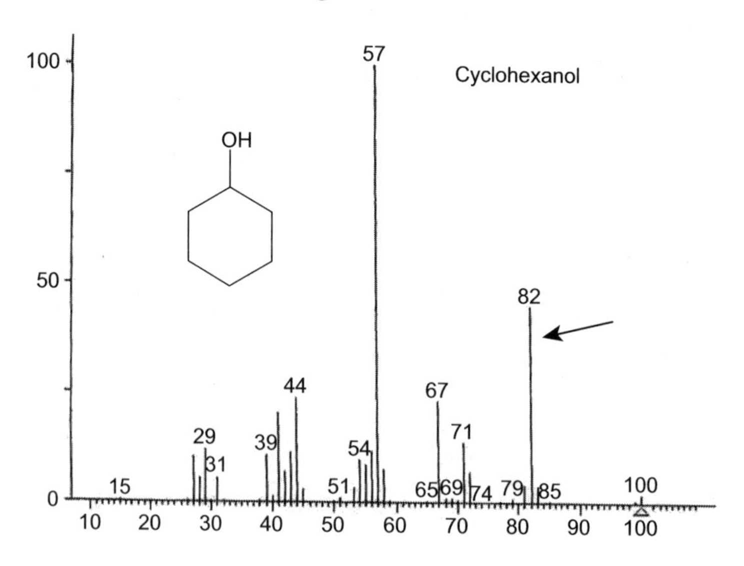

Figure 3.121

Answers to Selected Questions

Figure 3.2	Dodecane
Figure 3.6	Diethyl ether
Figure 3.7	Methyl propylether
Figure 3.8	Ethyl n-propylketone
Figure 3.9	Isopropyl alcohol
Figure 3.10	p-Bromotoluene
Figure 3.11	Ethoxybenzene
Figure 3.12	Butanoic acid
Figure 3.13	Anisole
Figure 3.14	Diethylbenzene
Figure 3.15	p-Methyl acetophenone
Figure 3.16	3-Octanol
Figure 3.17	1-Octanol
Figure 3.18	3-Hexanone
Figure 3.19	Toluene
Figure 3.20	Isopropyl benzene
Figure 3.21	N,N-dimethylbenzenamine
Figure 3.22	Ethyl benzoate
Figure 3.23	Bromobenzene
Figure 3.24	Aniline
Figure 3.25	p-Aminotoluene
Figure 3.26	1-Phenyl-1-butanone
Figure 3.27	n-Pentylisopropyl ether
Figure 3.28	Acetophenone
Figure 3.29	Benzaldehyde
Figure 3.30	2-Methyl-2-butanol
Figure 3.31	p-Methylbenzaldehyde
Figure 3.32	o-Chloronitrobenzene
Figure 3.33	(Z)-3,7-dimethyl-2,6-octadien-1-ol
Figure 3.34	Dibromodichloromethane
Figure 3.35	Epinephrine
Figure 3.36	Cyanogen bromide
Figure 3.37	Butanal
Figure 3.38	Trichloronitromethane

Suggested Readings and References

Books

1. Beynon J. H., R. A. Saunders, and A. F. Williams, *The Mass Spectra of Organic Molecules*, Elsevier Publishing Co., New York, 1968.
2. Biemann K., *Mass Spectrometry: Organics Chemical Application*, McGraw-Hill, New York, 1962.
3. Budzikewicz H., C. Djerassi, and D. H. Williams, *Mass Spectrometry of Organic Compounds*, Holden-Day, Inc., San Francisco, 1962.
4. Johnston R. A. W., *Mass Spectrometry Basics*, Cambridge University Press, London, 2003.
5. Lawson A. M., C. K. Lim, and W. Richmond, eds., *Current Developments in Clinical Application of HPLC, GC and MS*, Academic Press, London, 1980.
6. McFadden W., *Techniques of Combined Gas Chromatography and Mass Spectrometry*, John Wiley & Sons, New York, 1973.
7. Pavia D. L., G. M. Lampman, and G. S. Kriz, Jr., *Introduction to Spectroscopy*, Cengage Learning, 2013.
8. Silverstein R. M., G. C. Bassler, and T. C. Morrill, *Spectrometric Identification of Organic Compounds*, 7th ed., John Wiley, New York, 2014.
9. McLafferty F. W., *Interpretation of Mass Spectra*, 3rd ed., University Science Books, 1993.
10. Middleditch B. S., *Practical Mass Spectrometry, A Contemporary Introduction*, Plenum Press, New York, 1997.
11. Millard B. J., *Quantitative Mass Spectrometry*, Heyden & Son, London, 1978.
12. Milne G. W. A., ed., *Mass Spectrometry Techniques and Applications*, Wiley-Interscience, New York, 1971.

13. *FT-ICR/MS: Analytical Fourier Transform Ion Cyclotron Resonance Mass Spectrometry*, New York, VCR Publishers, 1991.
14. Jarvis K. E., *Handbook of Inductively Coupled Plasma Mass Spectrometry*, New York, Chapman and Hall, 1992.
15. *Ion Spectroscopies for Surface Analysis*, New York, Plenum Press, 1991.
16. *Analytical Microbiology Methods: Chromatography and Mass Spectrometry*, New York, Plenum Press, 1990.
17. *Continuous-Flow Fast Atom Bombardment Mass Spectrometry*, New York, John Wiley & Sons, 1990.
18. *Mass Spectrometry of Peptides*, Boca Raton, CRC Press, 1991.
19. Kostiainen R., *The Characterization and Analysis of Trichothecenes by Chemical Ionization and Tandem Mass Spectrometry*, Helsinki, Suomalainen Tiedeakatemia, 1989.
20. Adams R. P., *Identification of Essential Oils by Ion Trap Mass Spectroscopy*, San Diego, Academic Press, 1989.
21. *Liquid Chromatography/Mass Spectrometry: Techniques and Applications*, New York, Plenum Press, 1990.
22. Prokai L., *Field Desorption Mass Spectrometry*, New York, M. Dekker, 1990.
23. *Mass Spectrometry of Biological Materials*, New York, M. Dekker.
24. *Liquid Chromatography/Mass Spectrometry: Applications in Agricultural, Pharmaceutical, and Environmental Chemistry*, Washington, DC, American Chemical Society, 1990.
25. Marshall A. G., *Fourier Transforms in NMR, Optical, and Mass Spectrometry: A User's Handbook*, Amsterdam, New York, Elsevier, 1989.
26. *Novel Techniques in Fossil Fuel Mass Spectrometry*, Philadelphia, Pa., ASTM, 1989.
27. Lai S. T. F., *Gas Chromatography/ Mass Spectrometry Operation*, 1st ed., East Longmeadow, Mass., Realistic Systems, 1988.
28. *Lasers and Mass Spectrometry*, New York, Oxford University Press, 1990.
29. *Secondary Ion Mass Spectrometry: Principles and Applications*, Oxford, New York, Oxford University Press, 1989.
30. Busch, Kenneth L., *Mass Spectrometry: Techniques and Applications of Tandem Mass Spectrometry*, New York, VCH Publishers, 1988.
31. *The Analysis of Peptides and Proteins by Mass Spectrometry*, Chichester, (England), New York, John Wiley & Sons, 1988.

32. *Applications of Inductively Coupled Plasma Mass Spectrometry*, Glasgow (Scotland), New York, Chapman and Hall, 1989.
33. Raymond E., *Quadrupole Storage Mass Spectrometry*, New York, John Wiley & Sons, 1989.
34. *Analysis of Carbohydrates by GLC and MS*, Boca Raton, Fla., CRC Press, 1989.
35. Millard B. J., *Quantitative Mass Spectrometry*, Heyden & Son, London, 1978.
36. *Time-of-Flight Mass Spectrometry and its Applications*, Amsterdam; New York: Elsevier, 1994.
37. Pavia D. L., G. M. Lampman, and G. S. Kriz, Jr., *Introduction to Spectroscopy: A Guide for Students of Organic Chemistry*, W. B. Saunders Company, Philadelphia, 1979 Chapter 2.
38. Karasek F. W., *Basic Gas Chromatograph–Mass Spectrometry: Principles and Techniques,* Elsevier, New York, 1988.
39. *Fourier Transform Mass Spectrometry: Evolution, Innovation, and Applications*, Washington, DC, American Chemical Society, 1987.
40. *Mass Spectrometry in Biotechnological Process Analysis and Control*, New York, Plenum Press, 1987.
41. *Inorganic Mass Spectrometry*, New York, John Wiley & Sons, 1988.
42. McFadden W., *Techniques of Combined Gas Chromatography and Mass Spectrometry*, John Wiley & Sons, New York, 1973.
43. *Applications of Mass Spectrometry in Food Science*, Elsevier Applied Science, New York, 1987.
44. *Applications of New Mass Spectrometry Techniques in Pesticide Chemistry*, New York, John Wiley & Sons, 1987.
45. Benninghoven A., *Secondary Ion Mass Spectrometry: Basic Concepts, Instrumental Aspects, Applications, and Trends*, New York, John Wiley & Sons, 1987.
46. *Mass Spectrometry in Biomedical Research*, Chichester (England), New York, John Wiley & Sons, 1986.
47. White F. A., *Mass Spectrometry: Applications in Science and Engineering*, New York, John Wiley & Sons, 1986.
48. *Gas Chromatography/ Mass Spectrometry*, Berlin, New York, Springer-Verlag, 1986.
49. *Mass Spectrometric Characterization of Shale Oils: A Symposium*, Philadelphia, PA, ASTM, 1986.

50. *Gaseous Ion Chemistry and Mass Spectrometry*, New York, John Wiley & Sons, 1986.
51. Chapman J. R., *Practical Organic Mass Spectrometry*, Chichester (West Sussex); New York, John Wiley & Sons, 1985.
52. Duckworth H. E., *Mass Spectrometry*, 2nd ed. Cambridge, New York, Cambridge University Press, 1986.
53. Desiderio D. M., *Analysis of Neuropeptides by Liquid Chromatography and Mass Spectrometry*, Amsterdam, New York, Elsevier, 1984.
54. *Mass Spectrometry in Environmental Sciences*, New York, Plenum Press, 1985.
55. *Introduction to Mass Spectrometry*, New York, Raven Press, 1985.
56. *Analytical Pyrolysis: Techniques and Applications*, London, Boston: Butterworths, 1984.
57. Message G. M., *Practical Aspects of Gas Chromatography/Mass Spectrometry*, New York, John Wiley & Sons, 1984.
58. *Gas Chromatography / Mass Spectrometry Applications in Microbiology*, New York, Plenum Press, 1984.
59. *Tandem Mass Spectrometry*, New York, John Wiley & Sons, 1983.
60. Lawson A. M., C. K. Lim, and W. Richmond, eds., *Current Developments in Clinical Application of HPLC, GC and MS*, Academic Press, London, 1980.
61. Meuzelaar H. L. C., *Pyrolysis Mass Spectrometry of Recent and Fossil Biomaterials: Compendium and Atlas*, Amsterdam, New York, Elsevier Scientific Pub. Co., 1982.
62. Rose M. E., *Mass Spectrometry for Chemists and Biochemists*, Cambridge; New York, Cambridge University Press, 1982.
63. Goodman S. I., *Diagnosis of Organic Acidemias by Gas Chromatography Mass Spectrometry*, New York, A.R. Liss, 1981.
64. McLafferty F. W., *Interpretation of Mass Spectra*, 3rd ed. Mill Valley, Calif., University Science Books, 1980.
65. *Mass Spectrometry: Principles and Applications*, 2nd ed., New York, McGrawHill, 1981.
66. *Current Developments in the Clinical Applications of HPLC, GC, and MS*: Proceedings of a Symposium Held at the Clinical Research Centre, Watford Road, Harrow, Middlesex, UK, on May 30, 31, and June 1, 1979, London, New York, Academic Press, 1980.
67. Schlunegger U. P., *Advanced Mass Spectrometry: Applications in Organic and Analytical Chemistry*, 1st ed. Oxford, New York, Pergamon Press, 1980.

68. Middleditch B. S., *Mass Spectrometry of Priority Pollutants*, New York, Plenum Press, 1981.
69. Budde W. L., *Organics Analysis Using Gas Chromatography/Mass Spectrometry: A Techniques & Procedures Manual*, Ann Arbor, Mich., Ann Arbor Science Publishers, 1979.
70. Sccheinmann J., ed., *An Introduction to Spectroscopic Methods*, Chapter 2, Pergamon Press, New York, 1974.
71. *Time-of-Flight Mass Spectrometry*, (1st ed.) Oxford, New York, Pergamon Press, 1969.
72. *Practical Mass Spectrometry: A Contemporary Introduction*, New York, Plenum Press, 1979.
73. Scheinmann F., *An Introduction to Spectroscopic Methods for the Identification of Organic Compounds*, 1st ed., Oxford, New York, Pergamon Press, 2 volumes, (1970–74).
74. *Quadrupole Mass Spectrometry and its Applications*, Amsterdam, Elsevier Scientific Pub. Co.; New York, 1976.
75. Masada Y., *Analysis of Essential Oils by Gas Chromatography and Mass Spectrometry*, New York, John Wiley & Sons, 1976.
76. Gudzinowicz B. J., *Fundamentals of Integrated GC-MS*, New York, M. Dekker, 1976.
77. Wilson R. G., *Ion Mass Spectra*, New York, John Wiley & Sons, 1974.
78. Frigerio A., *Essential Aspects of Mass Spectrometry*, Flushing, N.Y., Spectrum Publications, distributed by Halsted Press, New York, 1974.
79. Williams D. H., *Principles of Organic Mass Spectrometry*, London, New York, McGraw-Hill, 1972.
80. Hill H. C., *Introduction to Mass Spectrometry*, 2nd ed. (revised by A. G. Loudo), London, New York, Heyden, 1972.
81. *Mass Spectrometry in Biochemistry and Medicine*, New York, Raven Press, 1974.
82. Safe S., *Mass Spectrometry of Pesticides and Pollutants*, CRC Press, 1973.
83. *Mass Spectrometry and NMR Spectroscopy in Pesticide Chemistry*, New York, Plenum Press, 1974.
84. McFadden W. H., *Techniques of Combined Gas Chromatography/Mass Spectrometry: Applications in Organic Analysis*, New York, John Wiley & Sons, 1973.
85. Johnston R. A. W., *Mass Spectrometry for Organic Chemists*, Cambridge University Press, London, 1972.

86. Melton C. E., *Principles of Mass Spectrometry and Negative Ions*, New York, M. Dekker, 1970.
87. *Metastable Ions*, Amsterdam, New York, Elsevier Scientific Pub. Co., 1973.
88. Hamming M. C., *Interpretation of Mass Spectra of Organic Compounds*, New York, Academic Press, 1972.
89. Johnstone R. A. W., *Mass Spectrometry for Organic Chemists*, Cambridge, University Press, 1972.
90. Maccoll A., *Mass Spectrometry*, London, Butterworths; Baltimore, University Park Press, 1972.
91. *The Fragmentation of Some Cyanohydrins and O-Acylcyanohydrins Upon Electron Impact*, Stockholm, Almqvist & Wiksell, 1968.
92. Porter Q. N., *Mass Spectrometry of Heterocyclic Compounds*, New York, John Wiley & Sons-Interscience 1971.
93. *Ion-Molecule Reactions*, New York, John Wiley & Sons-Interscience, 1970.
94. Knewstubb P. F., *Mass Spectrometry and Ion-Molecule Reactions*, London, Cambridge U. P., 1969.
95. Sharkey A. G., *Analytical Methods in Mass Spectrometry*, U.S. Department of the Interior, Bureau of Mines, (for sale by the Supt. of Docs., U.S. Govt. Print. Off.), 1967.
96. Budzikiewicz H., *Mass Spectrometry of Organic Compounds*, San Francisco, Holden-Day, 1967.
97. Beynon J. H., *The Mass Spectra of Organic Molecules*, Amsterdam, London, New York, Elsevier Pub. Co., 1968.
98. Roboz J., *Introduction to Mass Spectrometry; Instrumentation and Techniques*, New York, Interscience Publishers, 1968.
99. Budzikewicz H., C. Djerassi, and D. H. Williams, *Mass Spectrometry of Organic Compounds*, Holden-Day, Inc., San Francisco, 1967.
100. Ahearn A. J., *Mass Spectrometric Analysis of Solids*, Amsterdam, New York, Elsevier Pub. Co., 1966.
101. Blauth E. W., *Dynamic Mass Spectrometers*, Amsterdam, New York, Elsevier Pub. Co., 1966.
102. Reed R. I., *Applications of Mass Spectrometry to Organic Chemistry*, London, New York, Academic Press, 1966.
103. Roboz J., *Mass Spectrometry Instrument and Technique*, Interscience Publishers, New York, 1968.

104. Ramasastry J., *Mass Spectrometry: Theory and Applications*, New York, Plenum Press, 1966.

105. Middleditch B. S., *Practical Mass Spectrometry, A Contemporary Introduction*, Plenum Press, New York, 1979.

106. Pasto D. J. and C. R. Johnson, *Organic Structure Determination*, Prentice-Hall, Englewood Cliffs, N. J., 1969.

107. Kiser R. W., *Introduction to Mass Spectrometry and its Applications*, Englewood Cliffs, N.J., Prentice-Hall, 1965.

108. Budzikiewicz H., *Structure Elucidation of Natural Products by Mass Spectrometry*, San Francisco, Holden-Day, 1964.

109. Biemann K., *Mass Spectrometry: Organics Chemical Application*, McGraw-Hill, New York, 1962.

110. McDowell C. A., *Mass Spectrometry*, New York, McGraw-Hill, 1963.

111. Biemann K., *Mass Spectrometry: Organic Chemical Applications*, New York, McGraw-Hill, 1962.

112. Milne G. W. A., ed., *Mass Spectrometry Techniques and Applications*, Wiley-Interscience, New York, 1971.

113. Duckworth H. E., *Mass Spectroscopy*, Cambridge, University Press. 1958.

114. Pfleger K., *Mass Spectral and GC Data of Drugs, Poisons, Pesticides, Pollutants, and their Metabolites*, 2nd. rev., New York, VCH, 1992.

115. Beynon J. H., *Mass and Abundance Tables for Use in Mass Spectrometry*, Amsterdam, New York, Elsevier Pub. Co., 1963.

116. McLafferty F. W., *Mass Spectral Correlations*, Washington, American Chemical Society, 1963.

117. Lederberg J., *Computation of Molecular Formulas for Mass Spectrometry*, San Francisco, Holden-Day, 1964.

118. Beynon J. H., *Table On Meta-Stable Transitions for Use in Mass Spectrometry*, Amsterdam, New York, Elsevier Pub. Co., 1965.

119. McLafferty F. W., *Index and Bibliography of Mass Spectrometry*, 1963–1965 New York, Interscience Publishers, 1967.

120. ASTM Committee E-14 on Mass Spectrometry. Subcommittee IV on Data and Information Problems. *Index of Mass Spectral Data., Listed by Molecular Weight and the Six Strongest Peaks*, Philadelphia, American Society for Testing and Materials, 1969.

121. Neeter R., *Metastable Precursor Ions. A Table for use in Mass Spectrometry*, Amsterdam, New York, Elsevier Scientific Pub. Co., 1973.

122. Stenhagen E., *Registry of Mass Spectral Data*, New York, John Wiley & Sons, 1974.

123. Binks R., *Tables for Use in High Resolution Mass Spectrometry*, (London) Heyden in co-operation with Sadtler Research Laboratories (Philadelphia), 1970.

124. *Table of Ion Energies for Metastable Transitions in Mass Spectrometry*, Amsterdam, New York, Elsevier Pub. Co., 1970.

125. Mass Spectrometry Data Centre, *Eight Peak Index of Mass Spectra: The Eight Most Abundant Ions in 31,101 Mass Spectra*, Indexed by Molecular Weight, Elemental Composition and Most Abundant Ions, 2nd ed. Aldermaston, Mass Spectrometry Data Centre, 1974.

126. *Recent Developments in Mass Spectrometry in Biochemistry and Medicine*, International Symposium on Mass Spectrometry in Biochemistry and Medicine, Plenum Press, 1979.

127. *CRC Handbook of Mass Spectra of Drugs*, Boca Raton, Fla., CRC Press, 1981.

128. McLafferty F. W., *Mass Spectral Correlations*, 2nd ed., Washington, DC, American Chemical Society, 1982.

129. *Proceedings of the 9th International Mass Spectrometry Conference-Mass Spectrometry Advances*, Vienna, 30 August–3 September 1982, Amsterdam, New York, Elsevier Scientific Pub. Co., New York, N.Y., Distributors for the U.S. and Canada, Elsevier Science Pub. Co., 1983.

130. *An Eight Peak Index of Mass Spectra of Compounds of Forensic Interest*, Edinburgh, Published for the Forensic Science Society by the Scottish Academic Press, 1983.

131. Hites R. A., CRC *Handbook of Mass Spectra of Environmental Contaminants*, Boca Raton Fla., CRC Press, 1985.

132. Pfleger K., *Mass Spectral and GC Data of Drugs, Poisons, and Their Metabolites*, 1st ed., Weinheim, Federal Republic of Germany, VCH Verlagsgesellschaft; Deerfield Beach, FL, USA: Distribution, USA and Canada, VCH Publishers, 1985.

133. Beynon J. H., R. A. Saunders, and A. E. Williams, *The Mass Spectra of Organic Molecules*, Elsevier Publishing Co., New York, 1968.

Compilations of Mass Spectra

1. *Advances in Mass Spectrometry*, Heyden & Son, Ltd., London—7 Volumes from The Triannual International Conference on Mass Spectrometry.
2. *Biomedical Mass Spectrometry*, Heyden & Sons, Ltd., London.
3. CA Selects: *Mass Spectrometry*, Chemical Abstract Service, Columbus, Ohio.
4. Heller S. R. and G. W. A. Milne, *EPA/NIH Mass Spectral Data Base*, Obtained from the U.S. Government Printing Office.
5. *International Journal of Mass Spectrometry and Ion Physics*, Elsevier Scientific Publishers, The Netherlands.
6. *Mass Spectrometry Bulletin*, Mass Spectrometry Data Centre, Aldermaston, England.
7. *Mass Spectrometry*, Mass Spectrometry Society of Japan, Osaka, Japan (in English and Japanese)
8. *Journal of Mass spectrometry*, John Wiley and Sons.

Computer Programs and Mass Spectra Library Databases

1. *NIST/EPA/MSDC Mass Spectral Data Base*, Version 2.0, 1988, Copyright by the U.S. Secretary of Commerce and Distributed by the Office of Standard Reference Data of the National Institute of Standards and Technology. This database constitutes 276,248 mass spectra compounds.
2. WILEY Mass Spectra Library, 11th ed. This database constitutes 775,818 mass spectra.
3. *MASCOT* (Mass Spectrometry Calculations and Other Things), Fein-Marquart Associates. Inc., 7215 York Road, Baltimore, MD 21212 (301-821-5980). MS DOS version only. This program provides a number of helpful utilities for interpreting mass spectra data. Theoretical isotopic cluster patterns may be generated for comparison with actual spectra; formulas of fragment masses may be obtained from mass values; reference spectra may be searched and displayed. These are just a few of the many utilities afforded by the program.

Appendix

Table A.1 Neutral fragments expelled by multicentered fragmentation

Mass	Fragment	Mass	Fragment
2	H_2	45	C_2H_7N
17	NH_3	46	C_2H_6O or $H_2O + C_2H_4$
18	H_2O	48	CH_4S
20	HF	54	C_4H_6
27	HCN	56	C_4H_8 C_3H_4O
28	CO C_2H_4	58	C_3H_6O
30	CH_2O	59	C_3H_9N
31	CH_5N	60	C_3H_8O $C_2H_4O_2$
32	CH_4O	62	C_2H_6S
34	H_2S	74	$C_3H_6O_2$
36	HCl	76	C_6H_4
42	C_3H_6 C_2H_2O	78	C_6H_6
44	CO_2	80	HBr

Table A.2 Examples of some common fragment ions (this is not an exhaustive list)

m/z	Ion	*m/z*	Ion
15	CH_3^+	45	CH_3-O=CH_2, CO_2H^+
16	O^+	46	NO_2^+
17	OH^+	49	$CH_2=Cl^+$
18	H_2O^+, NH_4^+	51	$C_4H_3^+$
19	F^+	52	$C_4H_4^+$
27	$H_2C=CH^+$	61	$CH_3C(=OH)$ OH
28	CO^+, $CH_2=CH_2^+$	65	$C_5H_5^+$
	N_2^+(air)	69	CF_3^+, $C_5H_9^+$
29	$CH_3CH_2^+$, CHO^+	76	$C_6H_4^+$
30	$CH_2=NH_2^+$, NO^+	77	$C_6H_5^+$
31	$CH_2=OH^+$, CH_3O^+	78	$C_6H_6^+$
32	O_2^+ (air)	79	Br^+(also 81)
35	Cl^+(also 37)	91	tropylium ion ($C_7H_7^+$)
39	$C_3H_3^+$	91	$C_6H_5N^+$
41	H_2C=CH-CH_2^+	92	$C_6H_6N^+$
42	$CH_2=N=CH_2^+$	93	$C_6H_5O^+$
43	$C_3H_7^+$, $CH_3C=O^+$	105	Ph-C=O^+
44	CO_2^+	127	I^+

Note: Homologs of common fragment ions may be obtained by adding CH_2 to the above ions.

Source: F. W. McLafferty, *Interpretation of Mass Spectra*, 3rd ed., University Science Books (with some changes).

Table A.3 Some neutral fragments expelled by simple bond breaking

Mass	Fragment	Mass	Fragment
1	-H	47	$-CH_2SH$
15	$-CH_3$	49	$-CH_2Cl$
16	$-NH_2$	51	$-CHF_2$
17	-OH	54	$-CH_2CH_2CN$
19	-F	55	$-C_4H_7$ $-C_3H_3O$
26	-CN	57	$-C_4H_9$ $-C_3H_5O$
27	$-C_2H_3$	58	$-C_3H_8N$
29	$-C_2H_5$	59	$-C_3H_7O$ $-COOCH_3$ $-CH_2COOH$
30	-CHO $-CH_3NH$ $-CH_2NH_2$	61	$-C_2H_5S$
31	$-CH_3O$ $-CH_2OH$	65	$-C_5H_5$
33	-SH $-CH_2F$ $(H_2O + -CH_3)$	69	$-CF_3$
35	-Cl	71	$-C_4H_7O$
40	$-CH_2CN$	77	$-C_6H_5$
41	$-C_3H_5$	79	-Br $-C_6H_7$
43	$-C_3H_7$ $-CH_3CO$	81	$-C_6H_9$ $-C_5H_5O$
44	$-C_2H_5NH$	91	$-C_7H_7$
45	$-C_2H_5O$ -COOH	105	$-C_7H_5O$ $-C_8H_9$
46	$-NO_2$	121 127	$-C_7H_5O_2$ -I

Table A.4 Isotopic contributions of carbon

C#	$(X + 1)^+$	$(X + 2)^+$
C_1	1.1	0.00
C_2	2.2	0.01
C_3	3.3	0.04
C_4	4.4	0.07
C_5	5.5	0.12
C_6	6.6	0.18
C_7	7.7	0.25
C_8	8.8	0.34
C_9	9.9	0.44
C_{10}	11.0	0.54
C_{11}	12.1	0.67
C_{12}	13.2	0.80
C_{13}	14.3	0.94
C_{14}	15.4	1.1
C_{15}	16.5	1.3
C_{16}	17.6	1.5
C_{17}	18.7	1.7
C_{18}	19.8	1.9
C_{19}	20.9	2.1
C_{20}	22.0	2.3
C_{30}	33.0	5.2
C_{40}	44.0	9.4
C_{50}	55	15
C_{60}	66	21

Table A.5 Natural isotopic abundance and exact masses of common elements

Element	Symbol	Nominal mass	Exact mass	Abundance	X + 1 Factor	X + 2 Factor
Hydrogen	H	1	1.0078	99.985		
		2	2.0141	0.015		
Boron	B	10	10.0129	19.9	4.03nB	
		11	11.0093	80.1		
Carbon	C	12	12.0000	98.91		
		13	13.0034	1.1	1.1nC	0.0060(nC*nC)
Nitrogen	N	14	14.0031	99.63		
		15	15.0001	0.37	0.37nN	
Oxygen	O	16	15.9949	99.76		
		17	16.9991	0.04	0.04nO	
		18	17.9992	0.20		0.20nO
Fluorine	F	19	18.9984	100		
Silicon	Si	28	27.9769	92.2		
		29	28.9765	4.7	5.1nSi	
		30	29.9738	3.1		3.4nSi
Phosphorus	P	31	30.9738	100		
Sulfur	S	32	31.9721	95.0		
		33	32.9715	0.75	0.8nS	
		34	33.9679	4.22		4.4nS
Chlorine	Cl	35	34.9689	75.77		
		37	36.9659	24.23		32.5nCl
Bromine	Br	79	78.9183	50.7		
		81	80.9163	49.3		98.0nBr
Iodine	I	127	126.9045	100		

Note: Assume X = 100%: X represents the relative intensity of the first peak in a cluster of peak corresponding to isotopic variants of a given ion. Note that for all isotopic clusters, except Boron, the lowest mass isotope is the most abundant.
The factor is multiplied by the number (n) of atoms of the element in the ion to determine the X + 1 or X + 2 relative intensity contribution percentage of a given isotope. For example, the contribution at $m/z = X + 1$ due to 15N for an ion containing three nitrogen atoms would be $0.37 \times 3 = 1.11$ relative to 100 at $m/z = X$.

Table A.6 Relative intensity of peaks due to Cl/Br isotopes

ClBr	X	X + 2	X + 4	X + 6	X + 8	X + 10
Cl	100	32.5				
Cl_2	100	65.0	10.6			
Cl_3	100	97.5	31.7	3.4		
Cl_4	76.9	100	48.7	10.5	0.9	
Cl_5	61.5	100	65.0	21.1	3.4	0.2
Cl_6	51.2	100	81.2	35.2	8.5	1.1
ClBr	76.6	100	24.4			
Cl_2Br	61.4	100	45.6	6.6		
Cl_3Br	51.2	100	65.0	17.6	1.7	
$ClBr_2$	43.8	100	69.9	13.7		
Cl_2Br_2	38.3	100	89.7	31.9	3.9	
Cl_3Br_2	31.3	92.0	100	49.9	11.6	1.0
$ClBr_3$	26.1	85.1	100	48. 9	8.0	
Cl_2Br_3	20.4	73.3	100	63.8	18.7	2.0
Br	100	98.0				
Br_2	51.0	100	49.0			
Br_3	34.0	100	98.0	32.0		
Br_4	17.4	68.0	100	65.3	16.0	

Illustration: An ion with 2 chlorine atoms will have peaks at 2 and 4 m/z above the nominal mass with intensities of 65% and 10.6%, respectively, of that of the nominal mass peak.

Table A.7 Exact masses and natural isotopic abundance of elements

Element	Symbol	Nominal mass	Exact mass	Natural abundance
Hydrogen	H	1	1.007825	99.985
	D	2	2.01410x	0.015
Helium	He	4	4.00260	100
Lithium	Li	6	6.01512	7.5
		7	7.01600	92.5
Beryllium	Be	9	9.01218	100
Boron	B	10	10.0129	19.9
		11	11.0093	80.1

Table A.7 *Continued*

Element	Symbol	Nominal mass	Exact mass	Natural abundance
Carbon	C	12	12 (REF)	98.9
		13	13.00335	1.1
Nitrogen	N	14	14.00307	99.63
		15	15.00011x	0.37
Oxygen	O	16	15.99491	99.76
		17	16.99913	0.04
		18	17.99916	0.20
Fluorine	F	19	18.99840	100
Neon	Ne	20	19.99244	90.5
		21	20.9938	0.27
		22	21.99138	9.2
Sodium	Na	23	22.98977	100
Magnesium	Mg	24	23.9850	79.0
		25	24.9858	10.0
		26	25.9826	11.0
Aluminum	Al	27	26.9815	100
Silicon	Si	28	27.9769	92.23
		29	28.9765	4.67
		30	29.9738	3.10
Phosphorus	P	31	30.97376	100
Sulfur	S	32	31.9721	95.0
		33	32.9715	0.75
		34	33.9679	4.2
		36	35.9671	0.02
Chlorine	Cl	35	34.9689	75.8
		37	36.9659	24.2
Argon	Ar	36	35.9675	0.34
		38	37.9627	0.063
		40	39.9624	99.60
Potassium	K	39	38.9637	93.26
		40^{+}	39.9640	0.0117
		41	40.9618	6.73
Calcium	Ca	40	39.9626	96.94
		42	41.9586	0.647
		43	42.9588	0.135
		44	43.9555	2.09
		48	47.9525	0.187
Scandium	Sc	45	44.95591	100

Table A.7 *Continued*

Element	Symbol	Nominal mass	Exact mass	Natural abundance
Titanium	Ti	46	45.95263	8.0
		47	46.95180	7.3
		48	47.94795	73.8
		49	48.94787	5.5
		50	49.94480	5.4
Vanadium	V	50+	49.9472	0.250
		51	50.94396	99.750
Chromium	Cr	50	49.9461	4.35
		52	51.9405	83.79
		53	52.9407	9.50
		54	53.9389	2.36
Manganese	Mn	55	54.93805	100
Iron	Fe	54	53.9396	5.8
		56	55.9349	92
		57	56.9354	2.2
		58	57.9333	0.28
Nickel	Ni	58	57.9353	68.27
		60	59.9308	26.10
		61	60.9310	1.13
		62	61.9283	3.59
		64	63.9280	0.91
Copper	Cu	63	62.9296	69.17
		65	64.9278	30.83
Zinc	Zn	64	63.9291	48.6
		66	65.9260	27.9
		67	66.9271	4.1
		68	67.9248	19
		70	69.9253	0.6
Gallium	Ga	69	68.9256	60.1
		71	70.9247	39.9
Germanium	Ge	70	69.9242	20
		72	71.9221	27
		73	72.9234	7.8
		74	73.9212	36
		76	75.9214	7.8
Arsenic	As	75	74.9216	100

Table A.7 *Continued*

Element	Symbol	Nominal mass	Exact mass	Natural abundance
Selenium	Se	74	73.9225	0.9
		76	75.9192	9.0
		77	76.9199	7.6
		78	77.9173	24
		80	79.9165	50
		82	81.9167	9
Bromine	Br	79	78.9183	50.7
		81	80.9163	49.3
Krypton	Kr	78	77.9204	0.35
		80	79.9164	2.25
		82	81.9135	11.6
		83	82.9141	11.5
		84	83.9115	57
		86	85.9106	17.3
Rubidium	Rb	85	84.9118	72.17
		87 xx	86.9092	27.83
Strontium	Sr	84	83.9134	0.56
		86	85.9093	9.86
		87	86.9089	7.00
		88	87.9056	82.58
Yttrium	Y	89	88.90586	100
Zirconium	Zr	90	89.9047	51.45
		91	90.9056	11.22
		92	91.9050	17.15
		94	93.9063	17.38
		96	95.9082	2.80
Niobium	Nb	93	92.90638	100
Molybdenum	Mo	92	91.9068	14.8
		94	93.9047	9.25
		95	94.9058	15.9
		96	95.9047	16.7
		97	96.9058	9.55
		98	97.9054	24.1
		100	99.9076	9.63
Ruthenium	Ru	96	95.9076	5.5
		98	97.9055	1.9
		99	98.9061	12.7
		100	99.9030	12.6
		101	100.9056	17.0
		102	101.9043	31.6
		104	103.9054	18.7

Table A.7 *Continued*

Element	Symbol	Nominal mass	Exact mass	Natural abundance
Rhodium	Rh	103	102.90550	100
Palladium	Pd	102	101.9049	1.02
		104	103.9040	11.1
		105	104.9051	22.3
		106	105.9035	27.3
		108	107.9039	26.5
		110	109.9052	11.7
Silver	Ag	107	106.9051	51.84
		109	108.9048	48.16
Cadmium	Cd	106	105.9070	1.25
		108	107.9040	0.89
		110	109.9030	12.5
		111	110.9042	12.8
		112	111.9028	24.1
		113	112.9046	12.2
		114	113.9034	28.7
		116	115.9050	7.5
Indium	In	113	112.9041	4.3
		115^{+}	114.9039	95.7
Tin	Sn	112	111.9040	0.97
		114	113.9030	0.65
		115	114.9033	0.36
		116	115.9017	14.5
		117	116.9031	7.7
		118	117.9016	24.2
		119	118.9034	8.6
		120	119.9022	32.6
		122	121.9034	4.6
		124^{+}	123.9052	5.8
Antimony	Sb	121	120.9038	57
		123	122.9042	43
Tellurium	Te	120	119.9040	0.096
		122	121.9030	2.60
		123	122.9042	0.91
		124	123.9028	4.82
		125	124.9044	7.14
		126	125.9033	18.95
		128	127.9045	31.69
		130	129.9062	33.80
Iodine	I	127	126.90448	100

Table A.7 *Continued*

Element	Symbol	Nominal mass	Exact mass	Natural abundance
Xenon	Xe	124	123.9059	0.10
		126	125.9043	0.09
		128	127.9035	1.9
		129	128.9048	26
		130	129.9035	4.1
		131	130.9051	21
		132	131.9041	27
		134	133.9054	10.4
		136	135.9072	8.9
Cesium	Cs	133	132.90543	100
Barium	Ba	130	129.9063	0.106
		132	131.9050	0.101
		134	133.9043	2.42
		135	134.9050	6.59
		136	135.9044	7.9
		137	136.9058	11.2
		138	137.9052	71.7
Lanthanum	La	138^{+}	137.9071	0.09
		139	138.90610	99.91
Cerium	Ce	136	135.9071	0.19
		138	137.9060	0.25
		140	139.9054	88.5
		142	141.9092	11.1
Praseodymium	Pr	141	140.90736	100
Neodymium	Nd	142	141.9077	27.1
		143	142.9098	12.2
		144^{+}	143.9101	23.8
		145	144.9122	8.3
		146	145.9131	17.2
		148	147.9165	5.76
		150^{+}	149.9207	5.64
Samarium	Sm	144	143.9117	3.1
		147^{+}	146.9149	15.0
		148	147.9148	11.3
		149	148.9172	13.8
		150	149.9170	7.4
		152	151.9197	26.7
		154	153.9222	22.7
Europium	Eu	151	150.9199	48
		153	152.9212	52

Table A.7 *Continued*

Element	Symbol	Nominal mass	Exact mass	Natural abundance
Gadolinium	Gd	152+	151.9198	0.20
		154	153.9207	2.18
		155	154.9226	14.8
		156	155.9221	20.5
		157	156.9240	15.6
		158	157.9241	24.8
		160	159.9271	21.9
Terbium	Tb	159	158.92535	100
Dysprosium	Dy	156	155.9253	0.06
		158	157.9244	0.10
		160	159.9248	2.3
		161	160.9269	18.9
		162	161.9268	25.5
		163	162.9287	24.9
		164	163.9292	28.2
Holmium	Ho	165	164.93033	100
Erbium	Er	162	161.9288	0.14
		164	163.9293	1.61
		166	165.9303	33.6
		167	166.9321	22.9
		168	167.9324	26.8
		170	169.9355	14.9
Thulium	Tm	169	168.93423	100
Ytterbium	Yb	168	167.9339	0.13
		170	169.9349	3.1
		171	170.9363	14.3
		172	171.9364	21.9
		173	172.9382	16.1
		174	173.9389	32.
		176	175.9426	12.7
Lutetium	Lu	175	174.9408	97.41
		176+	175.9427	2.59
Hafnium	Hf	174	173.9400	0.162
		176	175.9414	5.21
		177	176.9432	18.61
		178	177.9437	27.30
		179	178.9458	13.63
		180	179.9466	35.10
Tantalum	Ta	180	179.9475	0.012
		181	180.94801	99.988

Table A.7 *Continued*

Element	Symbol	Nominal mass	Exact mass	Natural abundance
Tungsten	W	180	179.9467	0.1
		182	181.9482	26.3
		183	182.9502	14.3
		184	183.9510	30.7
		186	185.9544	28.6
Rhenium	Re	185	184.9530	37.40
		187	186.9558	62.60
Osmium	Os	184	183.9525	0.02
		186	185.9539	1.6
		187+	186.9560	1.6
		188	187.9559	13.3
		189	188.9582	16
		190	189.9585	26
		192	191.9615	41
Iridium	Ir	191	190.9606	37
		193	192.9629	63
Platinum	Pt	190	189.9599	0.01
		192	191.9614	0.8
		194	193.9627	33
		195	194.9648	34
		196	195.9649	25
		198	197.9679	7.2
Gold	Au	197	196.96656	100
Mercury	Hg	196	195.9658	0.1
		198	197.9668	10.0
		199	198.9683	16.8
		200	199.9683	23.1
		201	200.9703	13.2
		202	201.9706	29.8
		204	203.9735	6.8
Thallium	Tl	203	202.9723	29.52
		205	204.9744	70.48
Lead	Pb	204	203.9730	1.4
		206	205.9745	24.1
		207	206.9759	22.1
		208	207.9766	52.4
Bismuth	Bi	209	208.98039	100

+ Long lived radioactive isotopes.

x Frequently used in labeled compounds.

A.1 Graphic Presentation of Isotopic Peaks

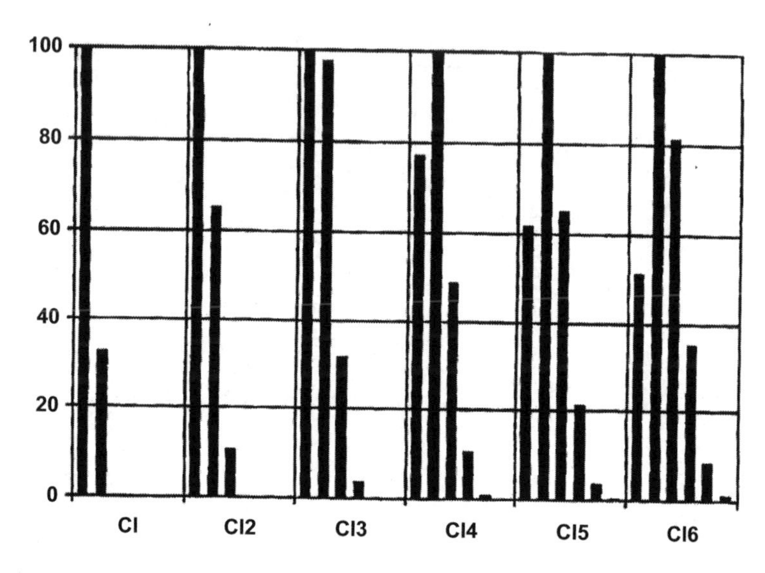

Figure A.1 Graphic presentation of isotopic peaks for Cl. . . Cl_6. Peaks are at intervals of 2 m/z (X, X + 2, X + 4, . . .).

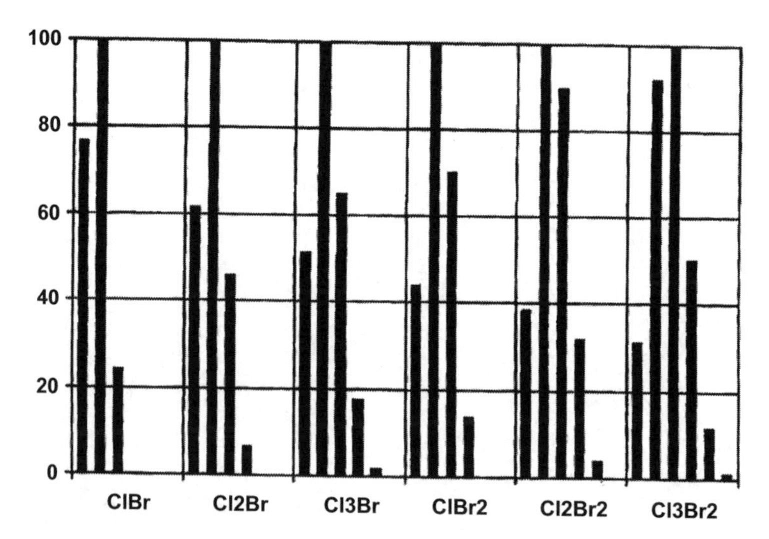

Figure A.2 Graphic presentation of isotopic peaks for combinations of chlorine and bromine: ClBr, Cl_2Br, Cl_3Br, $ClBr_2$, Cl_2Br_2, and Cl_3Br_2. Peaks are at intervals of 2 m/z (X, X + 2, X + 4, . . .).

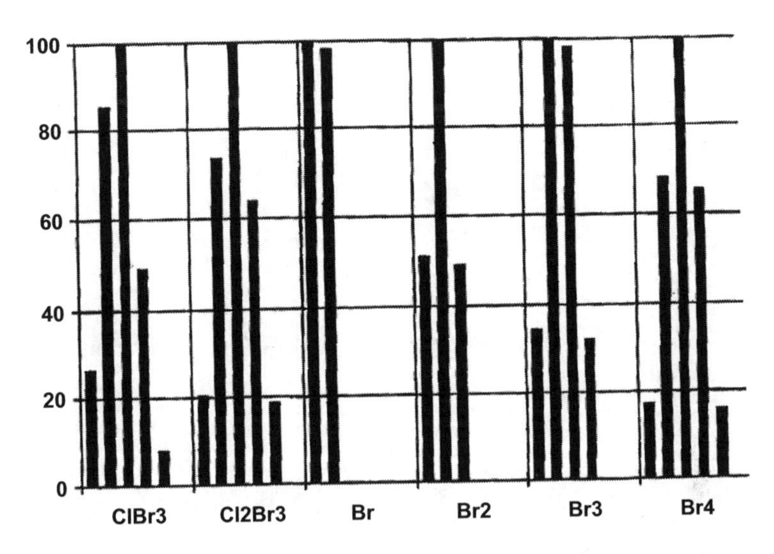

Figure A.3 Graphic presentation of isotopic peaks for BrCl Combinations: $ClBr_3$, Cl_2Br_3, Br, Br_2, Br_3, and Br_4. Peaks are at intervals of 2 m/z (X, X + 2, X + 4, . . .).

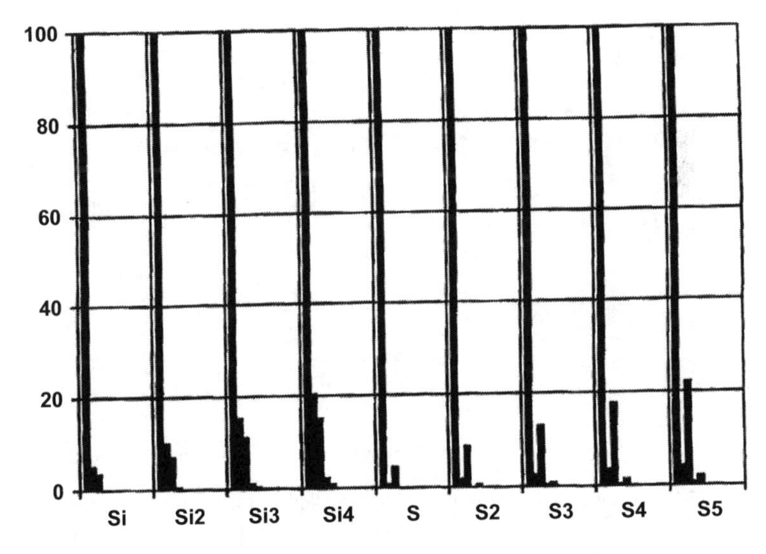

Figure A.4 Graphic presentation of silicon and sulfur isotopic peaks. Peaks are at intervals of 1 m/z (X, X + 1, X + 2, . . .).

Alkanes

a. Simple fission of C–C bonds, most frequently at the site of branching

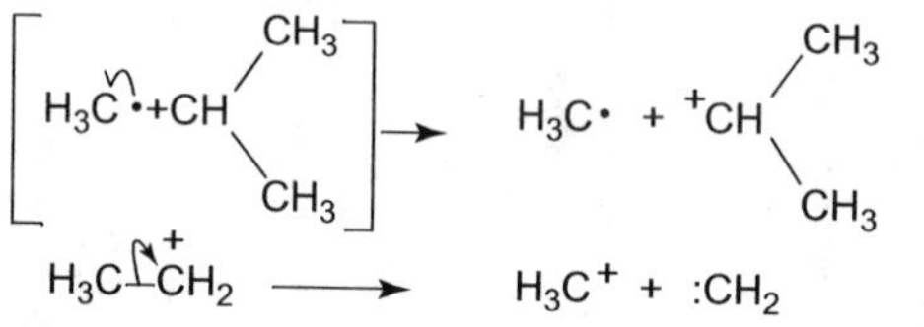

$$\left[H_3C\cdot{+}CH(CH_3)_2\right] \rightarrow H_3C\cdot + {}^{+}CH(CH_3)_2$$

$$H_3C\text{–}\overset{+}{C}H_2 \rightarrow H_3C^+ + :CH_2$$

b. Cyclic alkanes tend to lose side chains and/or extrude neutral olefinic moieties

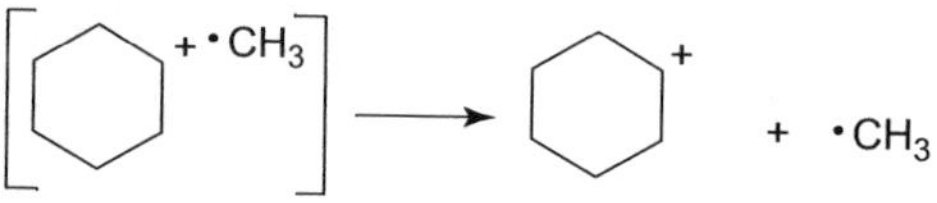

$$\left[C_6H_{11}\text{–}CH_3\right]^{+\cdot} \rightarrow C_6H_{11}^{+} + \cdot CH_3$$

Alkenes

a. Simple allylic cleavage (vinylic cleavage is much less frequent)

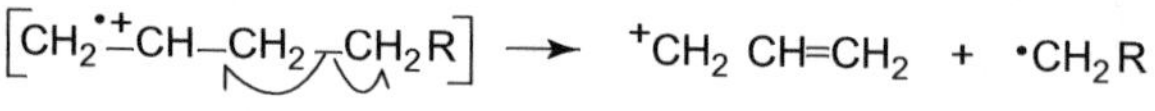

$$\left[CH_2\overset{\cdot+}{=}CH\text{–}CH_2\text{–}CH_2R\right] \rightarrow {}^{+}CH_2\ CH{=}CH_2 + \cdot CH_2R$$

b. McLafferty rearrangement (if γ H atoms are present)

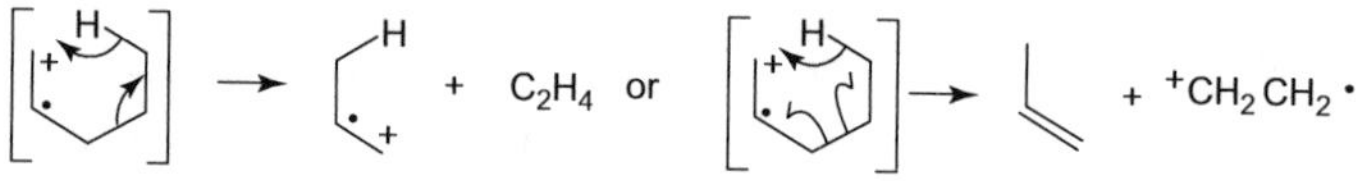

$$\rightarrow + C_2H_4 \text{ or } \rightarrow + {}^{+}CH_2CH_2\cdot$$

c. Retro-Diels Alder

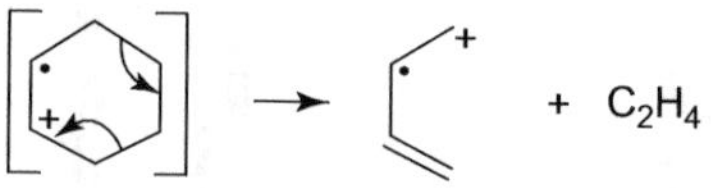

$$\rightarrow + C_2H_4$$

Figure A.5 Summary of fragmentation processes. Reprinted with permission from Creswell C. J., O. Runquist, and M. M. Campbell, *Spectral Analysis of Organic Compounds*, 2nd ed., Burgess Publishing Company, 1972.

Aromatic Hydrocarbons

a. Benzylic cleavage with ring expansion to the tropylium ion

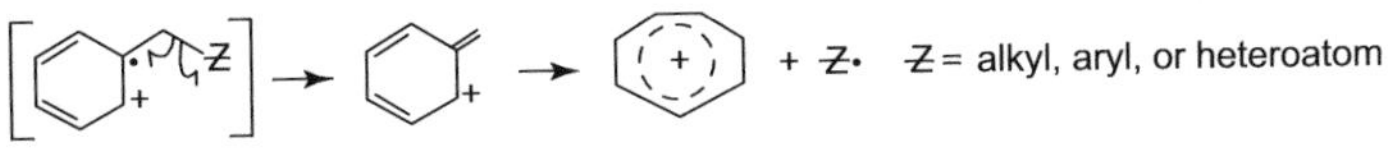

+ Z• Z = alkyl, aryl, or heteroatom

b. Vinylic cleavage

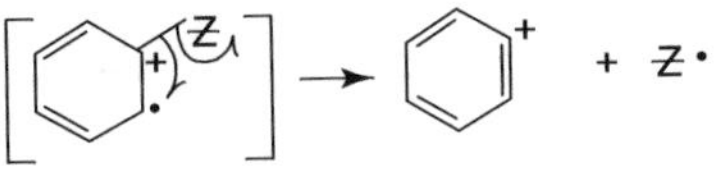

+ Z•

c. McLafferty rearrangement (if γ H atoms are present)

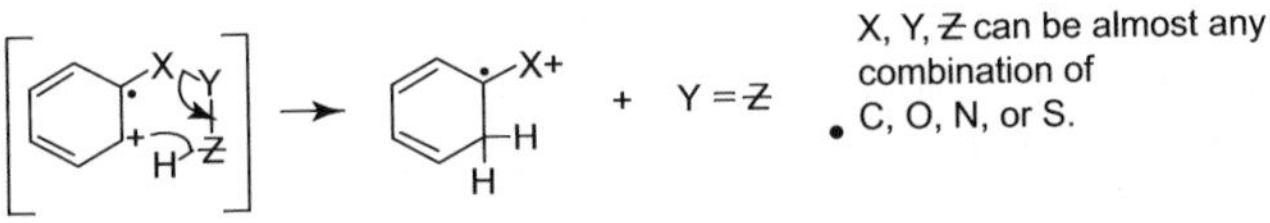

+ Y=Z

X, Y, Z can be almost any combination of C, O, N, or S.

d. Elimination of neutral fragments from *ortho*-disubstituted aromatic compounds

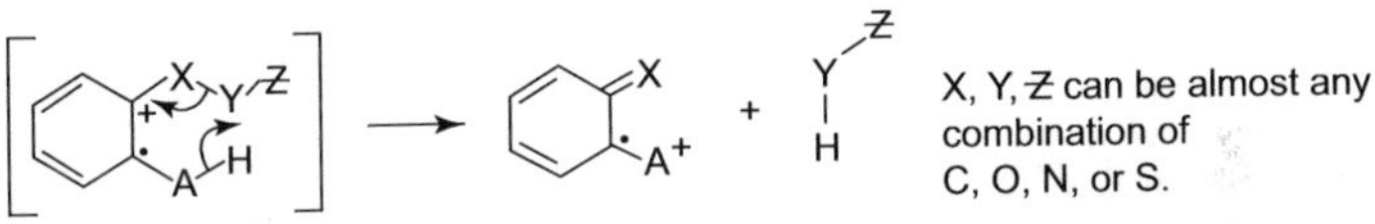

X, Y, Z can be almost any combination of C, O, N, or S.

e. Retro-Diels Alder

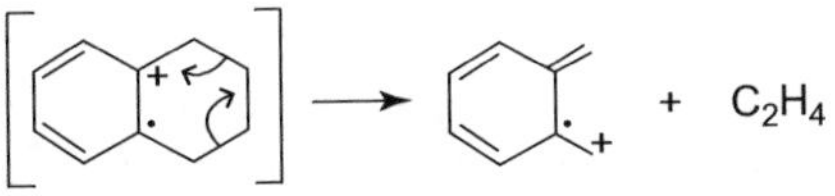

+ C_2H_4

Figure A.5 Continued.

Alcohols

a. Dehydration (thermal, prior to ionization and electron bombardment induced)

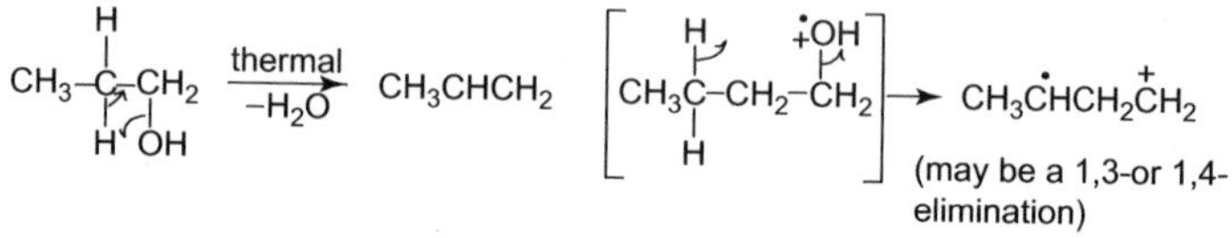

$CH_3-CH(H)-CH_2(OH) \xrightarrow[-H_2O]{thermal} CH_3CHCH_2$ $[CH_3C(H)_2-CH_2-\dot{C}H_2(\overset{\bullet}{+}OH)] \rightarrow CH_3\dot{C}HCH_2\overset{+}{C}H_2$

(may be a 1,3-or 1,4-elimination)

1,4-dehydration product may undergo further cleavage

$\rightarrow C_2H_4 + \dot{C}H_3$

b. α, β-cleavage to form oxonium ions

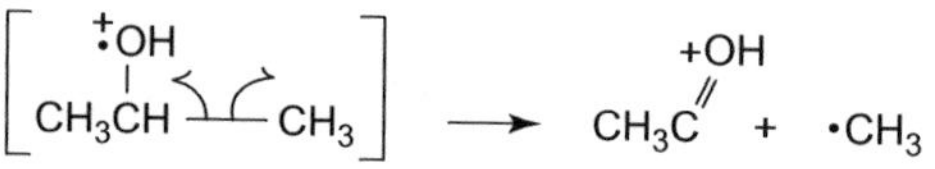

$[CH_3CH(\overset{+}{\bullet}OH)-CH_3] \rightarrow CH_3C(=\overset{+}{O}H) + \bullet CH_3$

c. Complex fission with H transfer in cyclic alcohols

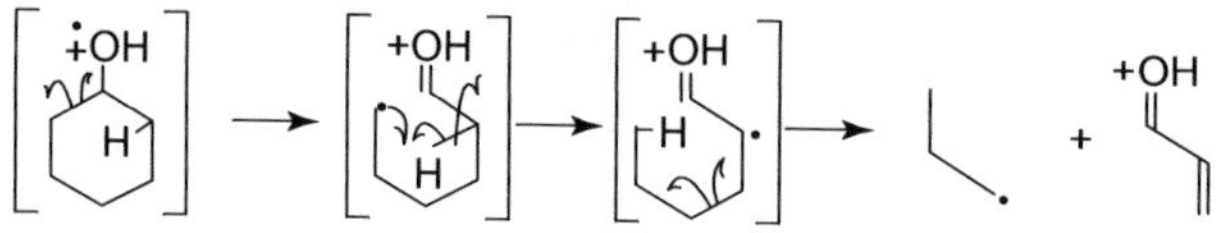

Aliphatic Amines

a. α, β-fission with formation of immonium ion

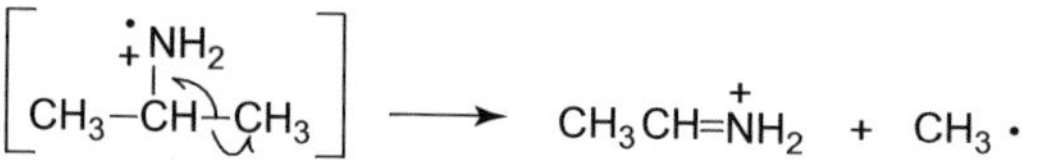

$[CH_3-CH(\overset{\bullet}{+}NH_2)-CH_3] \longrightarrow CH_3CH=\overset{+}{N}H_2 + CH_3\bullet$

Immonium ions may further cleave with transfer of H

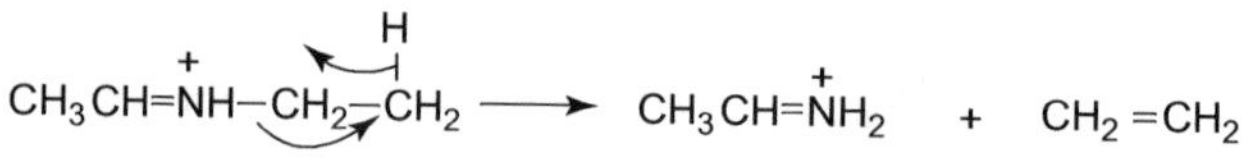

$CH_3CH=\overset{+}{N}H-CH_2-CH_2(H) \longrightarrow CH_3CH=\overset{+}{N}H_2 + CH_2=CH_2$

b. Complex fission with H transfer in hydrocarbon rings with amino substituents (analogous to complex fission of cyclic alcohols)

Figure A.5 Continued.

Aliphatic Ethers

a. Alkyl-oxygen fission. The charge usually resides with the alkyl moiety

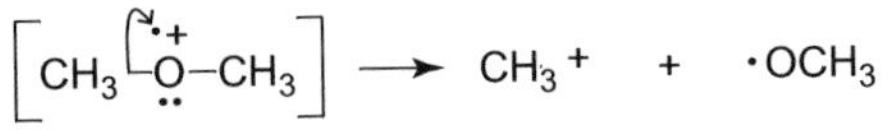

$$[CH_3-\dot{O}^{+}-CH_3] \rightarrow CH_3^+ + \cdot OCH_3$$

b. α, β-fission with oxonium ion formation

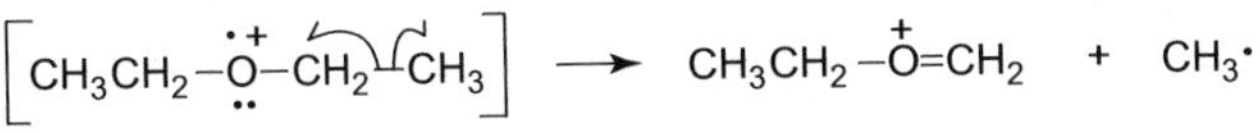

$$[CH_3CH_2-\dot{O}^{+}-CH_2-CH_3] \rightarrow CH_3CH_2-\overset{+}{O}=CH_2 + CH_3\cdot$$

The oxonium ion formed may undergo further cleavage with H transfer

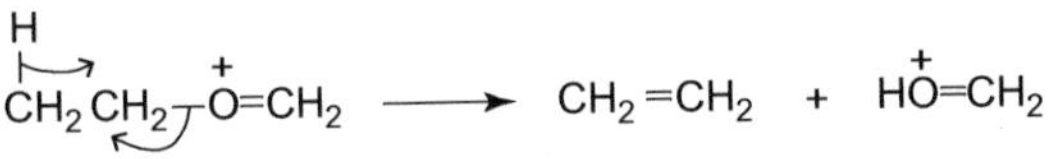

$$H-CH_2CH_2-\overset{+}{O}=CH_2 \rightarrow CH_2=CH_2 + H\overset{+}{O}=CH_2$$

c. Cyclic ethers may extrude a neutral aldehyde moiety

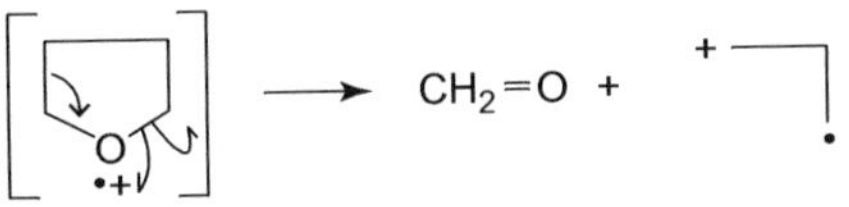

$$\rightarrow CH_2=O + $$

Halides

a. Cleavage of the H–X bond

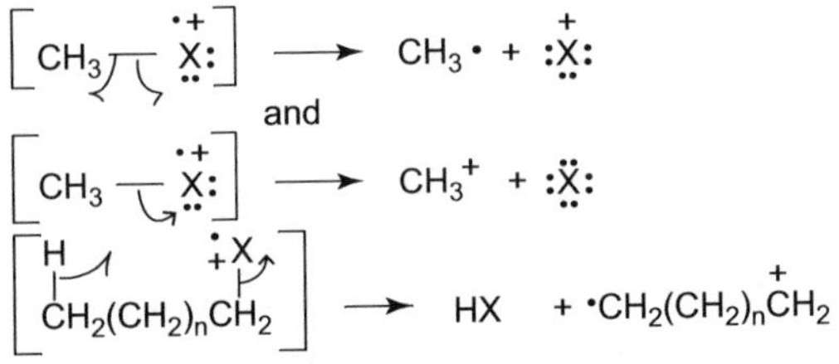

$$[CH_3-\dot{X}^{+}:] \rightarrow CH_3\cdot + :\overset{+}{X}:$$

and

$$[CH_3-\dot{X}^{+}:] \rightarrow CH_3^+ + :\ddot{X}:$$

$$[H-CH_2(CH_2)_nCH_2-\dot{X}^{+}] \rightarrow HX + \cdot CH_2(CH_2)_n\overset{+}{C}H_2$$

c. α, β-fission with the formation of halonium ion

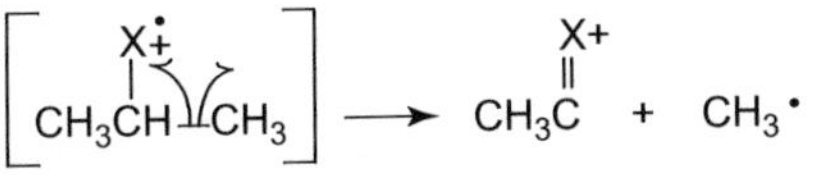

$$[CH_3CH(\dot{X}^{+})-CH_3] \rightarrow CH_3C(=X^+) + CH_3\cdot$$

Figure A.5 Continued.

d. Remote cleavage with the formation of cyclic halonium ions

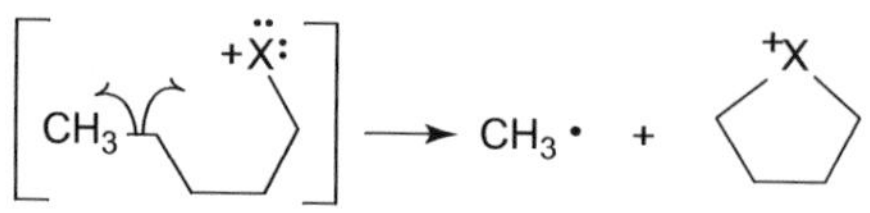

$[CH_3\text{–(ring)–}\ddot{X}^{+}:] \longrightarrow CH_3\bullet + \text{(cyclic } ^{+}X\text{)}$

Esters

a. α-cleavage to form ions of the type R^+, RCO^+, ^+OR, $^+OCOR'$ and R'^+

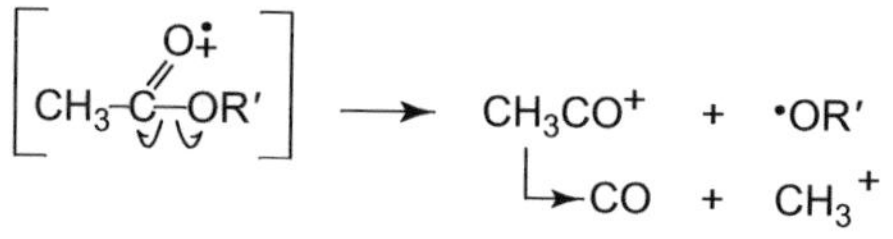

$[CH_3\text{–}C(=O^{\bullet+})\text{–}OR'] \longrightarrow CH_3CO^+ + \bullet OR'$

$CH_3CO^+ \longrightarrow CO + CH_3^+$

b. $[CH_3\text{–}C(=O^{\bullet+})\text{–}OR'] \longrightarrow CH_3^+ + \bullet OCOR'$

$[CH_3\text{–}C(=O^{\bullet+})\text{–}OR'] \longrightarrow \overset{+}{O}R' + CH_3CO\bullet$

$[CH_3\text{–}C(=O^{\bullet+})\text{–}O\text{–}R'] \longrightarrow CH_3\text{–}CO_2^{\bullet} + R'^+$

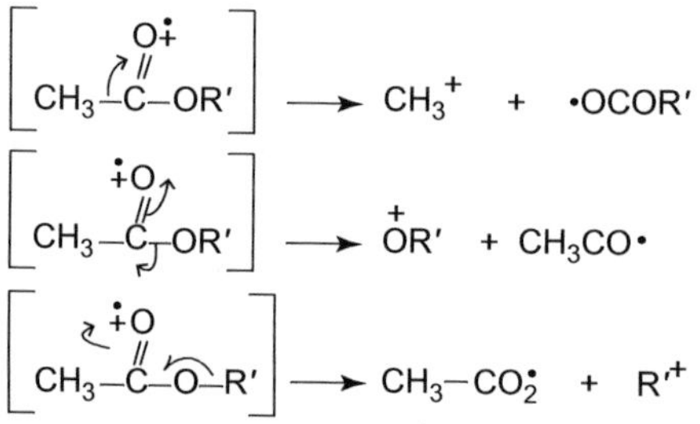

c. McLafferty rearrangement

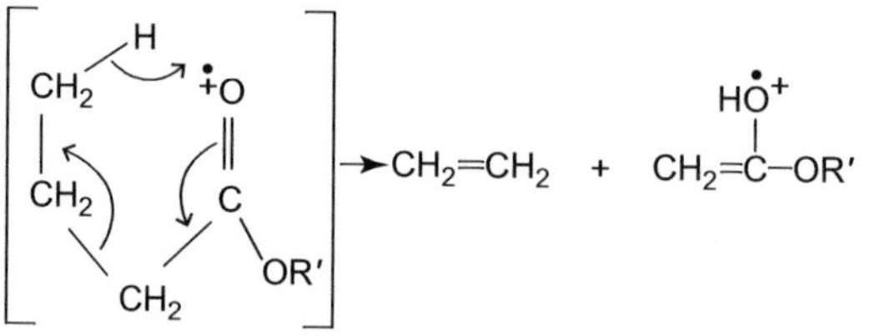

$[CH_2(H)\text{–}CH_2\text{–}CH_2\text{–}C(=O^{\bullet+})\text{–}OR'] \longrightarrow CH_2{=}CH_2 + CH_2{=}C(HO^{\bullet+})\text{–}OR'$

d. Double rearrangement of certain esters to yield protonated carboxylic acid fragments

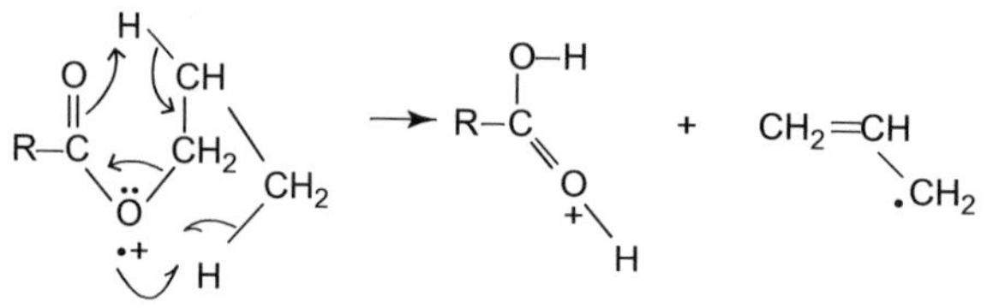

$[R\text{–}C(=O)\text{–}\ddot{O}^{\bullet+}\text{–}CH_2\text{–}CH(H)\text{–}CH_2(H)] \longrightarrow R\text{–}C(\text{–}O\text{–}H)(=\overset{+}{O}\text{–}H) + CH_2{=}CH\text{–}\dot{C}H_2$

Figure A.5 Continued.

Aldehydes and Ketones

a. α-cleavage to form ions of the type R^+ and RCO^+

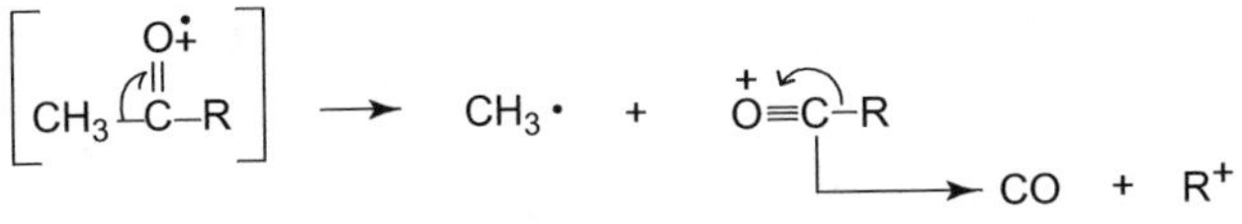

$$[CH_3-C(=O^{+\bullet})-R] \rightarrow CH_3^{\bullet} + \overset{+}{O}\equiv C-R \rightarrow CO + R^+$$

b. McLafferty rearrangement

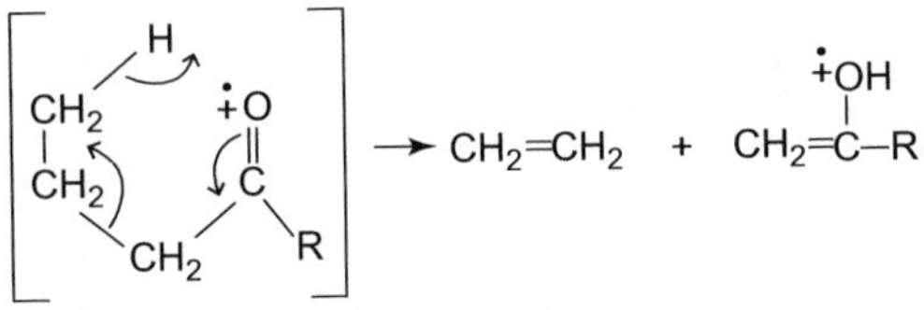

$$\rightarrow CH_2{=}CH_2 + CH_2{=}C(\overset{\bullet}{+}OH)-R$$

c. Cyclic ketones undergo complex fission to yield neutral fragments and an oxonium ion

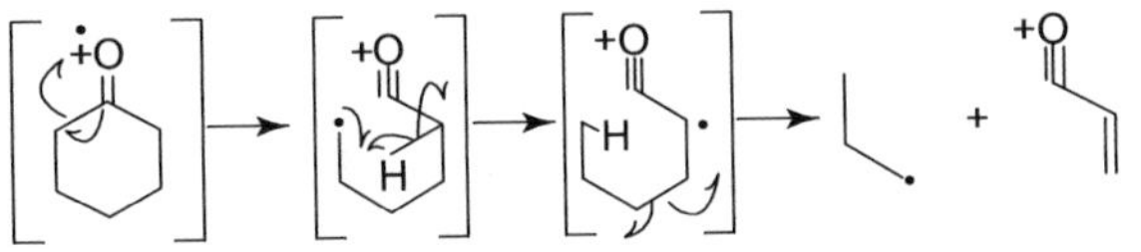

d. Bridged aromatic ketones extrude carbon monoxide

$$+ CO$$

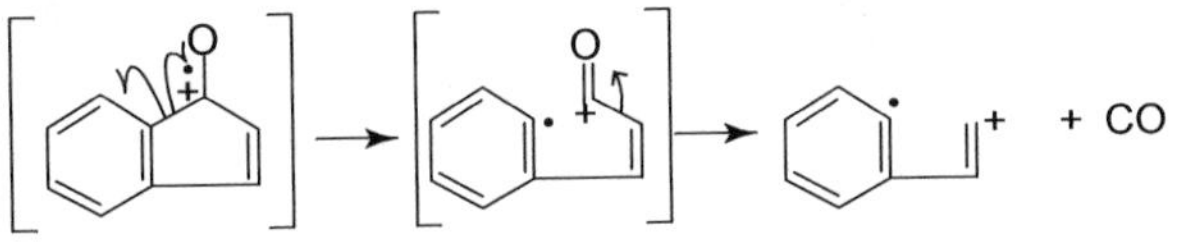

Phenols

a. Phenols extrude carbon monoxide

$$+ CO$$

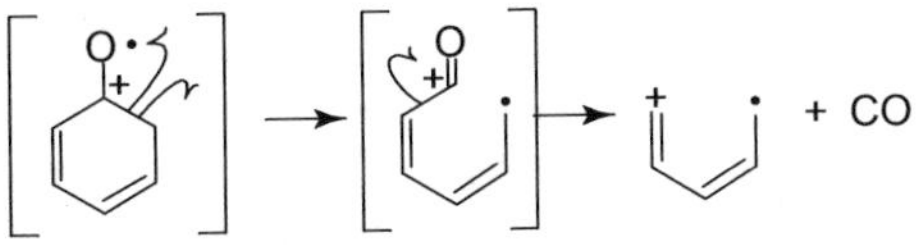

Figure A.5 Continued.

A.2 Silicon and Sulfur Isotopic Peaks

Peaks are at intervals of 1 m/z unit (X, X + 1, X + 2, . . .).

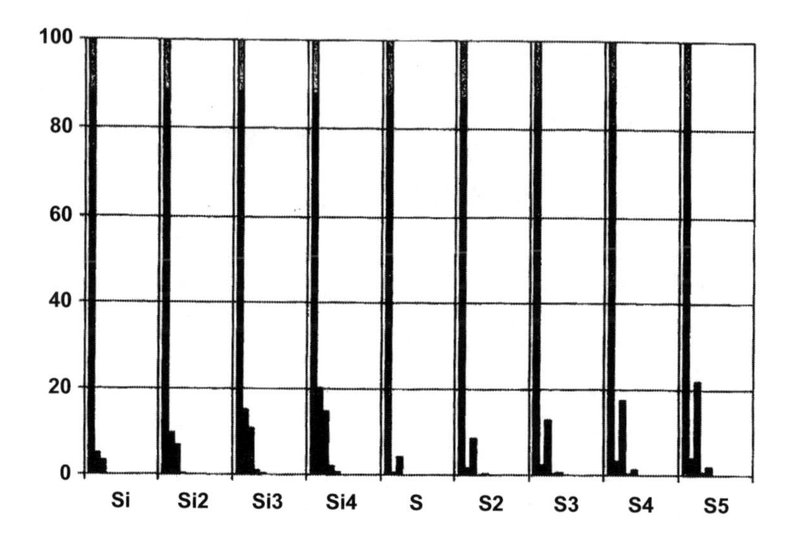

Index